The Fate of the Universe

THE FATE OF THE UNIVERSE

Richard Morris

PLAYBOY PRESS   NEW YORK

Manufactured in the United States of America.
FIRST EDITION
Playboy Press/A Division of PEI Books, Inc.

Library of Congress Cataloging in Publication Data

Morris, Richard, 1939–
  The fate of the universe.

  Bibliography: p.
  1. Cosmology.   I. Title.
QB981.M863   523.1   81-82462
ISBN 0-87223-748-6   AACR2

CONTENTS

ACKNOWLEDGMENTS

I would like to thank Drs. David Schramm and John Huchra for sending reprints and preprints of some of their papers. These proved to be indispensable in the writing of this book. Dr. Gustav Tammann was gracious enough to engage in a conversation with the author at a meeting of the American Association for the Advancement of Science. His remarks proved to be very illuminating. Dr. Jerome Kristian provided some very interesting astronomical photographs. Dr. E. N. Moore provided me with an opportunity to discuss some of this material in a colloquium at the University of Nevada. Some of the questions that were asked when I finished my presentation were very much to the point. And finally, my editor, Susan Ryan, was always ready to pounce on those passages which seemed a little murky.

The Fate of the Universe

CHAPTER 1

The Universe:
Open or Closed?

IN 1692, Sir Isaac Newton wrote a letter to the English clergyman and scholar Richard Bentley explaining why he believed the universe to be infinite. If the universe was finite, Newton said, then gravity would cause all of the matter in it to collect in its center. There would not be a multitude of individual stars, only a single huge mass. "But if the matter was evenly disposed throughout an infinite space," Newton went on, "it could never convene into one mass; but some of it would convene into one mass and some into another, so as to make an infinite number of great masses scattered at great distances from one to another throughout all that infinite space."

In an infinite universe, in other words, there would be no center. The countless stars and planets would maintain their positions because the average gravitational pull would be the same in every direction.

Apparently this argument did not convince Edmund Halley, the English astronomer after whom Halley's Comet was named. In 1720, Halley argued that an infinite universe was impossible. "If the number of Fixt Stars were more than finite," he said, "the whole superficies of their apparent Sphere would be luminous."

Halley thought that an infinite collection of stars would give off so much light that the night sky would be very bright. Wherever one looked, one would see a star. As a result, the sky at night—and by day as well—would be as bright as the sun. Today this argument is known as Olbers's paradox, after German physician and astronomer Heinrich Wilhelm Matthias Olbers, who resurrected it in 1826.

Newton and Halley were not the first to speculate on the question of an infinite or finite universe. Some of the early Greek philosophers, Anaximander and Democritus, for example, taught that there were an infinite number of worlds. Aristotle, who was skeptical of the doctrine, attempted to prove that both infinite space and infinite numbers were impossible. But neither side had the last word. The idea of infinite worlds is mentioned in the writings of the Roman poet Lucretius, and was one of the heresies for which the Italian philosopher Giordano Bruno was burned at the stake in 1600. Meanwhile, the more orthodox held with Aristotle. For example, in his *Divine Comedy* Dante described a universe which was finite and Aristotelian.

The concept of infinity has been creating problems ever since it was invented. The paradoxes associated with the idea are so numerous that a score of volumes the size of this book would not be sufficient to describe them all. And, as I will try to show in subsequent chapters, scientists are still troubled by the infinite; in physics, infinite quantities are often a sign that there is something wrong with one's theory.

But the finite Aristotelian universe was not without its paradoxes. In pre-Einsteinian days it seemed obvious that if the universe was finite, then it had to have an outer boundary. But what happens if one throws a ball beyond this boundary? Is the ball still part of the universe? It cannot have left the universe because the universe is, by definition, all that exists. On the other hand, it cannot still be in the universe because it has been thrown across the boundary.

When Newton and Halley discussed finite and infinite universes, they did not claim to have solved these ancient philosophical problems. They simply ignored them. Believing that the question of the "size" of the universe was one that science could answer, they appealed to scientific observation. They were more interested in what *was* than in what seemed philosophically reasonable.

Today we know that there were flaws in both Newton's and Halley's arguments; both began with assumptions that were not quite valid. Newton's argument erroneously assumed that if all matter in the universe were going to collect into one mass, it would already have done so. Newton was perfectly correct to assume that in a finite universe this would eventually happen. But he could not have known that this could be destined to happen billions of years in the future.

We know now that the universe is expanding. Galaxies

and clusters of galaxies are flying away from one another; the most distant ones are receding at velocities approaching the speed of light. Yet gravitational attraction between the galaxies is causing this expansion to slow down. If it slows down enough, the universe will eventually begin to contract.

Once the contraction begins, nothing will be able to stop it. Over a period of tens or hundreds of billions of years, the cosmos will get smaller and smaller. In time, everything will be compressed into a volume smaller than that occupied by the solar system—and the contraction will continue until the universe is literally crushed out of existence.

The flaw in Halley's argument is also related to the fact that the universe is expanding. Halley was correct to assume that in an infinite, static universe the sky would be as bright as the sun. But it is the expansion of the universe that dims the light from distant stars: A star that is rushing away from us does not look as bright as one that is approaching.

There is another reason why the night sky is black. The universe has a finite age; it was born in an enormous *big bang* 13 to 18 billion years ago. Now even in such a great length of time, a ray of light can travel only so far. Stars more than 13 to 18 billion light years away are invisible to us; the universe has not existed long enough for their light to reach us.

Who was right, Newton or Halley? Is the universe infinite or finite? This question has not yet been answered. Although our understanding of the universe has increased tremendously during the twentieth century, although numerous discoveries have been made in astronomy and cosmology, it is not yet possible to say what kind of universe we live in. The question is one of the great unsolved problems of contemporary science.

Today, scientists no longer speak of a finite or infinite cosmos; they speak instead of a universe that is *open* or *closed*. They use this terminology because it is more precise. According to Albert Einstein's theory of general relativity, there is an intimate connection between the "size" and the expansion of the universe.

In an open, or infinite, universe the expansion never stops; galaxies continue to recede from one another. The speed of recession may lessen, but things never come to a complete halt. Incidentally, there is no paradox here. When we say that an infinite universe is expanding, we mean simply that matter in it is becoming more dispersed. In a closed, finite universe, the expansion does eventually cease. And when it does, gravity begins to pull the galaxies together.

According to general relativity,* the answer to the question of whether the universe is open or closed depends on the mass in any given volume. If the average density of matter is more than a certain amount, then the universe is closed. If the density is less than that figure, then the universe is open. Naturally, the mass density changes with time. But for any given rate of expansion, it is possible to calculate a critical density. In other words, if we can deter-

---

* There are two theories of relativity. Einstein's theory of *special relativity*, published in 1905, deals with the behavior of bodies traveling at high velocities. *General relativity*, which was propounded in 1915, is Einstein's law of gravitation.

mine how rapidly the universe is expanding right now we can tell what mass density it must have now if it is closed.

The relationship between the amount of matter and the fate of the universe is a simple one. The greater the mass, the greater the gravitational force. The greater the gravitational force, the sooner the expansion slows down.

If the mass density is less than the critical amount, the gravitational brake will be weaker, and galaxies will continue to fly away from one another until they are out of gravity's range. As a result, the expansion will go on forever.

The relationship between matter density and the "size" of the universe is a complex one that depends upon the concept of *curved space*, which Einstein introduced when he published the theory of general relativity.

Curved space is not an easy thing to visualize. Fortunately it is possible to invent analogies. A very common one is that of a rubber sheet, the two-dimensional analogue of three-dimensional space. Imagine that the sheet is stretched on a frame so that it is perfectly flat. If a heavy weight is placed on the sheet, it will bend downward. The sheet will have been distorted in a way similar to the manner in which the presence of a large mass like the sun curves space in the Einsteinian universe. According to general relativity, the presence of matter curves space.

If a small spherical object is now rolled toward the depression caused by the weight, it will not travel in a straight line. Instead, it will be deflected as it rolls into the depression. If it is rolled in precisely the right way, it will go around in the depression in the way that a ball will roll around the rim of a roulette wheel. This is analogous to the revolution of a planet like the earth around the sun. Our small object will spiral in toward the middle of the depression (just as the roulette ball spirals in toward the center of the wheel and drops into a number slot), but this action is caused by friction. In empty space there is no friction, and a planet can circle a star forever.

Suppose that many weights are placed on the rubber sheet. Obviously they will have a cumulative effect; the entire sheet will bulge downward. If more weights are added, the bulge will be greater. In a similar way, all the objects in the universe contribute to the average curvature of space.

If the mass density is great enough, the curvature will be so great that space will close in upon itself. The two-dimensional analogue of such a closed universe would be a sphere. At this point our analogy breaks down; no one ever transformed a stretched rubber sheet into a sphere by piling weights upon it.

If the universe is closed, then it would be possible, in principle, to get into a spaceship and travel around the universe in more or less the way that Magellan circumnavigated the earth. Or at least one could do this if the universe did not collapse before one got all the way around. Unfortunately, Einstein's theory says it will collapse first.

One might think that if a closed universe resembles a sphere, it should be possible to get "outside" it. After all, airplanes and rockets rise above the two-dimensional surface of the earth. Here the analogy misleads us again. There is no "outside," because the universe contains all space and all time. Similarly, the universe does not expand "into" anything; space itself expands.

9

THE FATE OF THE UNIVERSE

The curvature of an open universe is of a different sort. Such a universe does not close in upon itself; it would contain the "infinite worlds" of Bruno and of the pre-Socratic philosophers Anaximander and Democritus, because the stars and galaxies in it go on forever.

It is possible to compare a closed universe to a sphere. The curvature of an open universe is sometimes compared to that of a saddle. However, one must not assume that because a sphere is a two-dimensional surface curved in three-dimensional space, our universe must have four spatial dimensions.

It is true that general relativity speaks of four-dimensional *space-time*. But in this sense Newton's laws are four dimensional too. Like those of Einstein, they also have three spatial coordinates and one time coordinate. But in general relativity, space and time interact with one another. In Newtonian physics they do not.

In other words, Einstein did not add a fourth dimension to our conception of space. He simply pointed out that it is not possible to talk of space as though it existed alone, and ignore the effects of time.

Because general relativity deals with matters that are so far removed from everyday experience, its implications often seem bizarre. Yet the theory is mathematically consistent; attempts to poke holes in it on purely logical or mathematical grounds are doomed to failure.

If a theory is consistent, it does not automatically follow that it must be true. If it is based on faulty assumptions, then its predictions will bear little relation to reality, no matter how logical or elegant it is. Only experimental tests and observations can tell us whether a theory is likely to be valid or not.

The Universe: Open or Closed?

Until fairly recently, the experimental evidence in favor of general relativity was minimal, and some scientists did entertain doubts about Einstein's ideas. Between 1915, when the theory was published, and the late 1960s, nearly a hundred alternative theories of gravitation were proposed. Physicists took some of them quite seriously, and there were times when it seemed that one theory or another might eventually replace Einstein's.

Today scientific journals rarely publish alternative theories, and acceptance of general relativity is practically universal. During the 1960s and 1970s it became possible to test the predictions of general relativity with unprecedented accuracy. In every case, when the evidence came in the verdict was in favor of Einstein.

This does not mean that general relativity provides the last word about the structure of the universe. There is no such thing as the ultimate scientific theory. Just as Newton's law of gravitation was superseded by Einstein's, it may be that Einstein's theory will have to be modified if scientists are to understand the nature of extremely intense gravitational fields, such as those that would exist in a contracting universe during the final stages of collapse.

But if this does turn out to be the case, it will still be possible to use general relativity when conditions are less extreme. Einstein's theory, like all of the "laws" of physics, is only an approximation. No theory ever describes nature exactly. But it is sufficient to describe the structure of the universe at the present time.

An analogous situation exists with respect to Newton's law of gravitation. Though this law is not exact, it is perfectly adequate for such tasks as calculating the orbits of space probes to Jupiter or Saturn. The use of general relativity for such a purpose is unnecessary; general relativity is mathematically more complicated, and the corrections it would provide are immeasurably small.

The relativistic corrections to Newton's theory are important only when gravity becomes very strong, or when one talks about the effects of gravity over distances of billions of light years. In most cases, the theory is just as usable as it was in Newton's day.

The corrections that a hypothetical "ideal" theory would make to relativity are smaller yet. It is likely that they would become important only under the first of the two sets of circumstances mentioned above: when gravitational forces become very intense. Since the gravitational fields that exist in interstellar space tend to be rather weak, we can be confident that general relativity gives an extremely accurate picture of the universe as a whole. Just as no new discovery is going to change the fact that the earth revolves around the sun, no new theory is likely to dramatically alter our conceptions of curved space.

But it has not always been possible to have such confidence. The early experimental tests of Einstein's theory were not very convincing, because general relativity gave results that differed only slightly from those predicted by Newton's law.

The first attempt to test general relativity was made less than four years after Einstein published the theory in the German scientific journal *Annalen der Physik* in 1915. In 1919, a team of scientists headed by the English astronomer Arthur (later Sir Arthur) Eddington went to the island of Principe, off the coast of Africa, to observe a solar eclipse. Meanwhile, another group of astronomers journeyed to Sobral, in Brazil. It was their intention to measure the deflection of starlight in the sun's gravitational field during a total eclipse.

According to Einstein's theory, the curvature of space near the sun should slightly bend a ray of starlight passing near the sun's surface. Unfortunately, rays of light that graze the sun's surface can only be seen during a total eclipse.

Eddington and his collaborators knew that the effect would be hard to measure. The curvature of space near the sun is very small, approximately one part in a million. However, a deflection was observed, and the results were widely heralded as confirmation of Einstein's theory. Einstein had predicted that the starlight would be deflected by 1.74 seconds of arc. The deflections measured by the Principe and Sobral expeditions were 1.61 and 1.98, respectively. Experimental uncertainties could account for the fact that they differed slightly from each other, and from Einstein's value, by a small amount. And if one averages the two experimental results, one obtains 1.80, which is even closer to the theoretical value.

It is true that 1.74 seconds of arc is a very small angle, only about 5/10,000 of one degree. Even today it is not easy to measure such small quantities accurately, especially under difficult conditions such as those attending an eclipse of the sun. Were Eddington's measurements accurate enough to be conclusive?

Since scientists do not like to leave such questions unanswered, the deflection of starlight was measured again during each subsequent total eclipse. However, a greater degree of accuracy was not achieved. In fact, the uncertainties grew larger. The results ranged from 1.43 seconds of arc to 2.7. It was obvious that starlight was being deflected, just as Einstein said it should be, but it was not certain that it was being deflected by exactly the right amount.

In 1915, one other test of the theory was possible—one based on the fact that the planet Mercury did not behave exactly as Newton's theory said it should. Like all the other planets, Mercury orbits the sun in a slightly elongated oval called an *ellipse*. And like all the other planets, when it is closest to the sun it is never at exactly the same place it was on the previous circuit.

The technical term for this is *precession of the perhelion. Perhelion* means "point of closest approach." The other planets in the solar system experience such precession to a lesser degree than does Mercury, because Mercury is closest to the sun.

The perhelion shifts by an amount of 1°33′20″ (1 degree, 33 minutes, and 20 seconds) per century. Of this, 1°32′37″ can be accounted for by Newton's theory; calculations show that the other planets should perturb Mercury's orbit by this amount. But the difference of 43 seconds of arc per century cannot be explained as due to Newtonian effects. During the latter part of the nineteenth century, this small discrepancy so troubled astronomers that they searched for a new planet inside Mercury's orbit, hoping it would be in the right place to provide the necessary perturbation. Of course the planet was never found.

General relativity predicted the shift correctly. Calculations showed that relativistic effects amounted to exactly

43 seconds of arc, and it appeared that Einstein's theory had triumphed.

Yet this confirmation of the theory left something to be desired. It was possible that the agreement between predictions and observations was accidental. For example, if the sun was not exactly spherical, the resulting anomalies in its gravitational field might produce the observed effect. And the orbit of Mercury was notoriously difficult to observe in the first place. Wasn't there some better way to test Einstein's theory? Other theories could be subjected to more precise tests, whose confirmation didn't depend on the resolution of minute discrepancies.

The problem was that Einstein's theory predicted deviations from the Newtonian law of gravity that could be measured only when gravity was very strong. Had scientists been able to travel to the surface of the sun, or at least so far out into space that the curvature became noticeable, they could have solved the problem.

By the early 1960s the situation had become critical. Numerous other gravitational theories were being proposed. Like general relativity, many of them also predicted such effects as the deflection of starlight by the sun, although in differing amounts. In 1971, Princeton University physicists Carl Brans and Robert Dicke published a theory of gravity that was briefly an especially formidable competitor to general relativity. For a while it seemed that the Brans-Dicke theory might explain some effects better than relativity did.

Even before the Brans-Dicke theory was proposed, it had become apparent that it was necessary to devise experimental tests that could distinguish between the effects predicted by general relativity and those predicted by the various other theories. Fortunately, a method for doing this became available at a very propitious time. In 1958 the German physicist Rudolf Mössbauer discovered a method for obtaining gamma rays of very precise frequencies from radioactive atoms embedded in crystals.

According to Einstein's theory, gravitational fields cause time to flow at a slightly slower rate. This is an effect that is related to the curvature of space; it is one of the ways in which time and space are bound up with one another. Now, gamma rays are a form of electromagnetic radiation; like radio waves and visible light, they have oscillation frequencies that can be measured in cycles per second. If gravity slows time down by a small amount, then it should cause the rate of oscillation of the gamma rays to change by the same amount. In other words, the gamma rays produced by the Mössbauer effect should change slightly as gravity becomes stronger or weaker.

Causing the strength of gravity to vary is quite simple; the gravitational field of the earth changes every time one climbs a flight of stairs. Because the second floor of a building is a little farther away from the center of the earth than the first, gravity is a little weaker there. The difference could never be measured with a bathroom scale, but is nevertheless very real.

In 1959 and 1960 Harvard physicist Robert V. Pound and his student Glen A. Rebka, Jr., made use of this fact by setting up an experiment in a 74-foot tower in Harvard's Jefferson Laboratory. The difference between the strength of gravity at the top of the tower and that at the bottom was very small, and the predicted frequency change was even more minute, only about one part in 10^{15} (ten to the fifteenth power, which is the numeral "1" followed by fifteen zeros—i.e., 1,000,000,000,000,000). However, since the Mössbauer effect made it possible to measure frequencies with this degree of precision, Pound and Rebka believed that a test of general relativity would be possible.

The experiment was successful. The results agreed with the effects predicted by relativity to an accuracy of about 1 percent. For the first time, a really convincing test of Einstein's theory had been made.

But scientists are never satisfied with one experiment. In order to be absolutely certain about their results, they perform experiments over and over again, in many different ways. If an unknown effect is influencing the results of one experiment, it is unlikely to crop up in all the others. Hence, even after Pound repeated the experiment with his colleague J. L. Snider in 1965, scientists still looked for ways to test general relativity.

In 1971, the experiment was repeated in a different way by Joseph C. Hafele, a physicist from Washington University in St. Louis, and Richard E. Keating of the U.S. Naval Observatory's Time Service. Hafele took extremely accurate atomic clocks aboard commercial jetliners and flew around the world with them. After the clocks had been compared with similar ones left on the ground, he boarded another plane and flew around the world in the opposite direction. If Pound and his collaborators had been able to measure changes in the rate of time flow at an altitude of 74 feet, Hafele and Keating reasoned, then even greater variations should be observable at the cruising altitudes of jet aircraft.

The clocks which Hafele and Keating used contained the isotope cesium 133, which emits microwaves* at a stable frequency of 9,192,631,770 cycles per second. Cesium clocks can measure time with an accuracy of 1 billionth of a second per day. Hafele and Keating were therefore able to measure the time variations predicted by general relativity to an accuracy of about 10 billionths of a second. Again, the agreement between theory and experiment appeared to be nearly perfect.

Since 1971 the experiment has been further improved. Atomic clocks have been placed in military aircraft and in rockets, and the time variations have been measured to an even greater accuracy. The results have demonstrated conclusively that as the gravitational fields change, so does the rate at which the clocks oscillate. The stronger the field is, the more time slows down.

Modern technology has also made it possible to measure other effects predicted by general relativity. According to the theory, a radar beam bounced off another planet should take slightly longer to travel from the earth to that planet and back again when its path grazes the sun. A radar beam, like a ray of light, will be deflected in the sun's gravitational field. A beam that follows a curved path will take slightly longer to get back to the earth than one which propagates in a straight line, and the difference should be measurable.

Many experiments designed to test this prediction have now been performed. Radar transmitters have been directed toward the planets Mercury, Venus, and Mars. In

---

* Microwaves are radio waves with wavelengths of less than one meter; they were so named because their wavelengths were shorter than those of the VHF band used by radar at the beginning of World War II.

every case, the return of the radar echo is delayed slightly when the sun is between the earth and the planet being studied.

Eddington's experiment of 1919 has been improved upon. One of the difficulties the English astronomer faced arose from the fact that the deflection of starlight could be measured only during an eclipse of the sun. Another problem was that it was necessary to compare the apparent position of the star during the eclipse to previous observations of its position in the sky. Things would have been much easier if Eddington had been able to see a star shift its position as the sun approached it. Unfortunately, this was not possible.

Experiments that surmounted these problems were finally performed, on a number of different occasions, during the early 1970s. Astronomers looked, not at stars, but at radio waves emitted by distant objects called *quasars*. Quasars, which are believed to be the luminous cores of distant galaxies, are located billions of light years from the earth. The fact that they are located at such great distances gives them the appearance of pinpoints of light. When they are observed through telescopes, they look like stars. As a result, their positions can be determined very exactly.

Most quasars emit radio waves. Since the sun is a weak radio source, it doesn't blot out radio emissions the way it does starlight. Hence quasars can be observed with radio telescopes at any time of the day or night, without the interference of the sun. If one wants to see whether the sun will bend the path of radio waves, it is not necessary to wait for an eclipse.

Every October 8, the sun passes in front of quasar 3C 273 (the label indicates that it is object number 273 in the Third Cambridge Catalogue), bending the light and radio waves as they travel from the quasar to the earth. Furthermore, 3C 273 is very near another quasar, called 3C 279. On October 8, 1972, radio astronomers monitored the separation between the two quasars as 3C 273 approached the sun. If the predictions of general relativity were correct, the angle between the two objects would change slightly. The radio waves from 3C 273 would be bent, while those from 3C 279 would continue to travel in a straight line. The astronomers found that this is exactly what happened. Einstein's theory had been confirmed again.

Other tests have been made. In 1974, radio astronomers using the Arecibo radio telescope in Puerto Rico discovered a *binary pulsar* about 15,000 light years away. A pulsar is the burned-out, collapsed relic of a star which emits pulses of light or radio waves at regular intervals; a binary pulsar is one that is a member of a double star system. This particular object, named Pulsar 1913 + 16 after the coordinates of its position in the sky, emits flashes of radio energy seventeen times per second. These flashes make it possible to perform very precise measurements of Pulsar 1913 + 16's orbit around its companion star, even though the companion shines too dimly to be seen from the earth. These shifts are analogous to the precession of the orbit of the planet Mercury. The only difference is that the perhelion of 1913 + 16 precesses at a much faster rate because it and its companion are very close. This makes the gravitational effects stronger. Hence it is possible to test general relativity to a very great degree of precision. Again, the agreement is very good.

This does not exhaust the list of the tests of general relativity that have been made since Pound and Rebka carried out their experiment in 1959. There have also been studies of the effect of gravity on light emitted by the sun, and observations of shifts in the orbits of planets other than Mercury. Laser beams have been aimed at pieces of reflecting material left on the moon by astronauts, making it possible to measure both the moon's orbit and the effects of general relativity to an accuracy of inches. Scientists have looked for slight variations in the earth's rotation rate, and they have done experiments to see whether, contrary to Einstein's theory, the strength of gravity might vary with time.

In every case the predictions of general relativity were confirmed. And then in 1979 astronomers discovered an effect that they hadn't even been looking for, one which even Einstein had thought impossible to observe: They discovered a *gravitational lens.*

In a 1937 article Einstein had mentioned the possibility that the image of a distant star could be split in two by the gravitational field of an intervening one. If both stars were positioned in just the right way, the light from the more distant star would bend around both sides of the nearer one. One would then see two images of the distant star. However, Einstein had concluded that "there is no great chance of observing this phenomenon."

The reason why Einstein did not think that a gravitational lens effect could be seen was that no known stars were massive enough to produce measurable splitting. Though a galaxy would be massive enough, until the late 1970s, most astronomers did not take the gravitational-lens idea very seriously, and there was no concerted effort to look for one.

THE FATE OF THE UNIVERSE

Early in 1979, Dennis Walsh of the Jodrell Bank radio astronomy observatory in England, Robert F. Carswell of Cambridge University, and Ray J. Weymann of the University of Arizona discovered two quasars that were extremely close to one another. The separation was so small that the quasars quickly became known simply as "the twins," rather than by their official designations, 0957 + 561 A and 0957 + 561 B (as in the case of the previously mentioned binary quasar, the numbers indicate the coordinates of the objects in the sky). Working with the 2.1-meter telescope at the Kitt Peak National Observatory and the 2.3-meter telescope at the University of Arizona, Walsh, Carswell, and Weymann were able to determine that the quasars were practically identical in every respect. They appeared to be equally bright, they were the same distance from the earth, and breakdowns of the light they emitted indicated that the same wavelengths were present. The three astronomers suggested, therefore, that the double image might be caused by a gravitational-lens effect, that the twins might not be two quasars, but rather a double image of one.

Further observations by groups of astronomers from the Massachusetts Institute of Technology, the University of Hawaii, and the California Institute of Technology seemed to shoot down this hypothesis. Slight inequalities between the two quasars were found, which seemed to prove that this was not a case of a split image.

But then astronomers Peter Young, James E. Gunn, Jerome Kristian, J. Beverley Oke, and James A. Westphal of the Hale Observatories and Alan N. Stockton of the University of Hawaii at Manoa proved that a gravitational-lens effect was being seen. A large galaxy surrounded by a cluster of smaller galaxies was found to lie between the two quasar images. The galaxy had not been seen before because it was so far away that under the best of conditions it produced only a blob of slight fuzz on astronomers' photographic plates.

The galaxy cluster did not lie exactly between the two quasar images; instead, it had an off-center position. So the Hale astronomers calculated, in rigorous mathematical detail, the projected effects of such an off-center position. They found that it should produce exactly the inequalities that had been observed. The hypothesis that the two quasars were one had been confirmed.

There is no reason why the gravitational-lens effect cannot produce more than two images. In fact, one would expect this to happen when the object that causes the splitting is an extended body such as a galaxy. In 1980, one such multiple splitting was observed. In June of that year, a triple quasar, PG1115 + 08, was reported. Further observations showed that the quasar had, not three images, but five. It seemed that the image designated A was really made up of three images superimposed one on another. PG1115 + 08 consisted, in other words, of one triple image and two single ones.

The observation of a triple or quintuple quasar strengthens the case for the gravitational-lens effect. It would be possible for a skeptic to argue that a double quasar was the result, not of a gravitational lens, but of some kind of remarkable coincidence. Although this is highly unlikely, it would be difficult to prove that our hypothetical skeptic was wrong. But a quintuple quasar is much harder to explain away; the skeptic is now required to account for four extra images, rather than just one.

So much evidence has been accumulated that no one can doubt the validity of the general theory of relativity. In fact only one important effect predicted by the theory has not yet been detected: This is the existence of *gravitational radiation* or, to use a slightly simpler term, *gravity waves*. They have not been detected because the existing experimental apparatus is not sensitive enough.

Gravity waves are feeble ripples in the curvature of space that are caused by massive bodies. That they should exist was pointed out by Einstein in 1916. Unfortunately, in most cases gravity waves are so weak that there is no hope of ever finding them. One must look for waves created by violent events, such as the supernova explosions in which massive stars are blown apart, and even then it is not easy.

This is not particularly surprising; of all the forces of nature, gravity is by far the weakest. For example, it is 10^{37} times weaker than electromagnetism, the force that produces such effects as magnetism, electricity, and radio waves. This is demonstrated every time a magnet is used to pick up an object like a nail: As small as the magnet is, it is able to overcome the gravitational force exerted by the entire earth.

An example of the minute effects that gravity waves are expected to have is provided by the American physicists Charles Misner, Kip Thorne, and John Archibald Wheeler in their textbook *Gravitation*. Misner, Thorne, and Wheeler calculate the effect that a supernova in the constellation Virgo would have on a typical piece of experimental apparatus. They conclude that it would be necessary for the experiment to measure a displacement of 3×10^{-19} (3 *divided by* 10^{19}) centimeters, approximately a billionth of a ten-billionth of an inch.

The situation improves slightly if the supernova takes place in a galaxy that is closer—the Virgo cluster is about 40 million light years away, which is relatively close. And it is probably hopeless to wait for a supernova explosion to occur in our own galaxy, the Milky Way. The last observed supernova took place before the invention of the telescope.

It is possible that gravity waves may already have been detected indirectly. In 1978 a team of radio astronomers studying a binary pulsar announced their observation of an energy change equal to the amount of energy that gravity waves should have carried away.

The pulsar, PSR1913 + 16, consists of a pair of objects which revolve around one another about once every 7.75 hours. Late in 1978, at a symposium held in Munich, Joseph Taylor, Peter McCulloch, and Lee Fowler of the University of Massachusetts announced that the period of revolution was decreasing by 0.0001 second every year.

They were not able to demonstrate that gravity waves were the only possible cause. Nevertheless, the case for the existence of gravitational radiation was certainly strengthened when they reported their work.

At this point, lest the reader think that I have strayed too far from the question of an infinite or finite universe, I will summarize the main points of this chapter:

1. Einstein's theory of general relativity tells us that there is a connection between the fate of the universe and its "size." If the universe is infinite, or open, then the present expansion will go on forever. On the other hand, if the universe is finite, or closed, the expansion will eventually halt, and the universe will enter into a phase of contraction.

2. It has become very difficult to doubt the validity of

general relativity. Between 1959 and 1980, Einstein's theory was subjected to numerous and varied tests. It triumphed in every case.

3. In order to tell whether the universe is open or closed, it is necessary to find out—directly or indirectly—the density of its matter. If the mass density is above a certain critical figure, then it must be closed. If the mass density is below this figure, it is open. Subsequent chapters will describe the attempts that have been made to measure this mass density.

CHAPTER 2

Redshifts

AFTER he had published the general theory of relativity, Einstein sought solutions to his equations. His laws of gravitation showed that matter curved space, but only additional calculations were needed to see what this implied about the universe as a whole.

As Einstein worked on the problem, he realized that his theory implied that the universe was expanding. At the time, everyone assumed that the universe was static. In 1916, most astronomers still believed that the Milky Way galaxy *was* the universe.

Believing that he had to find a solution which would correspond to a static universe, Einstein looked around for a way to "fix up" his equations. There was apparently only one way to do this, and Einstein reluctantly took the step.

THE FATE OF THE UNIVERSE

In 1917, he published a paper which described his conception of the cosmos.

Einstein's universe was closed. The assumption that there was no expansion or contraction led him to the conclusion that space was finite, that it closed in upon itself.

In order to obtain a static universe, Einstein added a term to his equations which he called the *cosmological constant*. The cosmological constant represented a kind of antigravity force, one that would balance out gravitational attraction at large distances. Unlike all other forces known to physics, it apparently increased with distance, rather than growing weaker.

At first glance, it seemed that Einstein had come up with a mathematical demonstration that the universe was closed, one that would have been very convincing if it were not for the presence of that odd cosmological fudge factor. Not only was the force it represented very strange, it was a force that had never been observed.

Years later, Einstein was to repudiate his 1917 paper, calling the introduction of the cosmological constant "the greatest blunder of my life." For in 1929 the American astronomer Edwin Hubble announced his discovery that the universe was not static at all; on the contrary, it was in a state of rapid expansion. Einstein realized that he could have predicted this expansion if he had only believed in his own theory and resisted the temptation to fix up the equations.

Hubble's discovery would dramatically alter the accepted concept of the universe. In 1917, astronomers did not understand the structure of the Milky Way; they did not know that the earth and the sun were not in the galactic center, but rather three-fifths of the way to the rim. Nor did they realize that there were galaxies beyond the Milky Way.

Other galaxies had been seen, but their distances from earth could not be measured. Consequently no one realized how large they were. Most astronomers erroneously believed that they were glowing patches of interstellar gas within the Milky Way, and referred to them as *nebulae*, the Latin word for clouds.

At the time, this belief was perfectly reasonable, given the paucity of observational evidence. The telescopes of the day were not powerful enough to distinguish individual stars within distant galaxies; astronomers knew only that they were looking at fuzzy patches in the sky. The only clusters that they could resolve into stars were comparatively close and contained relatively few members.

A hundred and fifty years had passed since the German philosopher Immanuel Kant had suggested that some nebulae, those with oval or spiral shapes, were really "island universes" that lay at enormous distances. Unfortunately, no compelling scientific evidence existed to support Kant's hypothesis. Astronomers conjectured that the spirals were whirlpools of gas in which new stars were being formed. The observation of a supernova in the Andromeda galaxy in 1885 had seemed only to confirm this theory. At the time, scientists did not realize that supernovae were not "new stars" at all, but massive dying stars being blown apart in violent explosions. Nor did they know how intrinsically bright supernovae were.

They were not even bothered by the fact that the supernova of 1885 disappeared after shining brightly for a short time. The new star, they concluded, was obviously going through a kind of settling-down process; it would undoubtedly flare up again.

It often seems that scientific revolutions occur just when scientists become most complacent about their theories. In 1917 one such revolution was brewing. Although most astronomers did not realize it, the first step toward the measurement of galactic distances had been taken five years before. Soon they would be forced to admit that some of their nebulae were millions of light years away, and that they really were "island universes," comparable in size to the Milky Way.

Most astronomical distances are measured by comparing an object's apparent brightness to its intrinsic luminosity. Stars and galaxies become dimmer when they are farther away. If we look at two stars or two galaxies of the same type and find that one is much brighter than the other, we can conclude that the brighter one is closer. The problem, of course, lies in determining the intrinsic brightness. For example, it would be possible to tell how far away a light bulb was by measuring its apparent brightness, but only if one knew beforehand that it was 25 or 60 or 100 watts.

In 1912, Henrietta Leavitt, a Harvard College astronomer working in South Africa, made a discovery that was to foment an astronomical revolution. Of course Leavitt had no idea that she was doing any such thing; she was only trying to accumulate some useful information about variable stars.

Variables are stars that pulsate. Their luminosity increases, decreases, and then increases again. In some the changes in light output are very regular; in others they are not. Perhaps the best known regular variable is Polaris, the

North Star, which reaches a peak of brightness once every 3.97 days.

Today it is known that the changes in brightness of variable stars are caused by cycles of expansion and contraction. Naturally, in 1912 Leavitt knew nothing about these changes in size. But she did notice that in certain types, called *Cepheid variables*, there was a relationship between period and luminosity. The brighter a Cepheid was at its peak, the longer the period between peaks.

The Cepheids that Leavitt studied were in the Small Magellanic Cloud, which can be seen only in the Southern Hemisphere (there is also a Large Magellanic Cloud; they are named after the Portuguese navigator Ferdinand Magellan, who observed them during his voyage around the world in the sixteenth century). At the time it was not known that the Cloud was a small satellite galaxy that revolved around the Milky Way, but it was recognized that all of the stars in it were about the same distance from the earth.

This last fact enabled Leavitt to compare one Cepheid to another. If one looked twice as bright, then it was intrinsically twice as bright. If it appeared to be dimmer, then it had a lower intrinsic luminosity.

Leavitt's discovery meant that these variables could be used to calculate distances in the universe. If astronomers could measure the distance to just one Cepheid, then the period–luminosity relation could be used to find the distances to them all. And with these figures, it would be possible to compute the distance to any collection of stars that included a Cepheid among its members.

By the time Edwin Hubble came to California's Mount Wilson observatory in 1919 to study the nebulae with the new 100-inch telescope, the distances of a few nearby Cepheids had been determined. At last, astronomers had a tool that could be used to measure large distances. Hubble used this tool to prove that some of the nebulae were really other galaxies.

During the early 1920s, Hubble discovered Cepheids in the nebulae known as NGC 6822 (number 6822 in the New General Catalogue), M31 (number 31 in the Messier catalogue, named after Charles Messier, a nineteenth-century French astronomer), and M33. By measuring the brightness of the Cepheids, Hubble was able to show that these nebulae were hundreds of thousands of light years away. This meant that they could not possibly lie within the Milky Way. And since they were collections of numerous stars rather than patches of gas, they had to be very large. Kant's island universes really existed after all.

Hubble was also able to compare the distance of each galaxy with its speed of recession. It had previously been discovered that light from most of the galaxies was shifted toward the red, indicating that they were moving away from the earth. Hubble was able to show that the amount of the redshift was proportional to the distance. This indicated that the entire universe was in a state of rapid expansion.

It is not hard to see why light from receding objects should be redshifted. Light is a wave phenomenon. Hence a ray of light can be thought of as a series of crests and troughs, analogous to the crests and troughs of ocean waves.

If an object is moving away, each successive wave crest must travel a longer distance before it reaches us. For example, imagine an automobile that beeps at 1-second intervals, and is moving away at a speed of 30 feet per second. If the first beep must travel 100 feet before it reaches us, then the second must travel 130, the third 160, and so on.

If, on the other hand, an object is moving toward us, the crests will be bunched together. In this case, the wavelength (distance from crest to crest) will be shortened.

In the case of light, wavelength is correlated with color; red light has the longest wavelength, while that of blue light is much shorter. Since wavelengths become longer when an object recedes, the light it emits will be reddened. On the other hand, light from an approaching object will shift toward the blue.

Light from distant galaxies does not look red to the eye, because galaxies also emit invisible ultraviolet "light." As the visible part of the spectrum is reddened, some of the ultraviolet shifts into the visible blue wavelengths. The net effect, as far as the eye is concerned, is that nothing has happened; the galaxy looks the same as it did before.

But the galaxy will look very different to scientific instruments that can split light up into its component wavelengths. These instruments, called *spectrographs*, can measure the shift of any wavelength. Spectrographs can also be used to determine the chemical composition of stars and galaxies. One need only determine what wavelengths are present to ascertain what chemical elements are present in the body that emitted the light.

The fact that light from the great majority of the galaxies was redshifted meant that they were flying away from the earth. But Hubble realized that this did not imply that the earth was in the center of the expansion, or that it was standing still while everything else was moving. The galaxies seemed to be moving away from the earth because they all were flying away from one another. An expanding universe has the same appearance wherever one happens to be; there is no center.

Two analogies can be used to illustrate this point: Imagine that a number of dots are painted on the surface of a balloon, and that the balloon is then blown up. Each dot will recede from every other dot. Or think of some raisin-bread dough that has just been put into an oven. As the dough expands (i.e., rises), the distances between the raisins increase.

Furthermore, the dots or raisins that are farthest from one another will increase their distance the fastest. This is exactly analogous to the state of affairs in the universe; the galaxies farthest from us recede at the highest velocities. Finally, the fact that there is a relationship between distance and velocity allows us to calculate how rapid the expansion is.

The only complication relates to the fact that a few galaxies exhibit blueshifts. One of these is the great galaxy in Andromeda, known to astronomers as M31. M31 is approaching the Milky Way.

The statement that all galaxies recede from one another is actually an oversimplification. Galaxies come in clusters that are held together by gravitational forces. It is the clusters, not the individual galaxies, that recede from one another.

The Milky Way and M31 are both members of a cluster of galaxies known as the *Local Group*. Light from other Local Group galaxies can be either redshifted or blueshifted, depending upon their relative motion. A redshift in the Local Group is unrelated to the expansion of the universe.

The fact that galaxies come in clusters makes life easier for astronomers; if the redshift for one galaxy in a distant cluster can be measured then it is the same for all the others. This can be quite a simplification, for some clusters contain thousands of members.

Hubble announced his discovery that the universe was expanding in 1929. During the next few years Hubble and his collaborator Milton Humanson (who, incidentally, was initially employed by the Mount Wilson observatory as a mule packer; he worked his way up to the position of astronomer after a number of years) observed progressively more distant galaxies. Cepheid variables could be distinguished only in the nearest galaxies, but once their distances had been measured, it was possible to calculate the luminosity of their brightest stars. Since the bright stars in different galaxies were very much alike, they could be used as distance yardsticks in place of the Cepheids; the principle was exactly the same. Hubble and Humanson simply looked for similar luminous stars in different galaxies, and measured their apparent brightness.

Hubble's discovery of the expansion of the universe showed that Einstein's paper on the static universe had been based on a mistaken assumption. Yet Hubble also concluded that the universe was closed. In his 1936 book, *The Realm of the Nebulae,* he offered the opinion that "if the red shifts are velocity shifts, it follows that the universe is closed, having finite volume and finite contents."

Hubble had found small discrepancies in his redshift data which seemed to indicate that the universe had to be closed. Hubble thought he had determined that the number of galaxies per unit volume changed with distance. That is, if one looked out into the universe a distance of, say, 100 million light years, the galaxies would not be as numerous as they were in the neighborhood of the Milky Way.

But it quickly became obvious that, like Einstein, Hubble had come to his conclusion too hastily. In 1936, it was impossible to make measurements accurate enough to support this belief. Today, scientists believe that Hubble's result was based on a systematic measuring error.

Furthermore, Hubble had based his conclusion on the existence of small discrepancies in his data. But it soon became apparent that Hubble's results contained a much larger discrepancy.

The problem was Hubble's figure for the age of the universe. If the universe is expanding, then it is a simple matter to extrapolate backward and to estimate the amount of time that has passed since the expansion began. After all, if everything is flying apart, there must have been a time when all matter in the universe was packed tightly together. Even if one is not sure how quickly the expansion is slowing down, it is still possible to obtain an upper limit on the age of the universe, called the *Hubble time.*

According to Hubble's measurement, the universe could not be more than 2 billion years old. But in 1936, it was already known that certain terrestrial rocks were 3.5 billion years old. Radioactive dating had established that fact beyond any reasonable doubt. Obviously, rocks on the earth could not be older than the universe itself.

Einstein and Hubble agreed that the universe was closed. But neither scientist could offer convincing evidence. One had concluded this from the erroneous assumption that the universe was static, the other from an expanding-universe theory that looked as though it contained an enormous flaw.

Was the universe really closed? No one knew. By 1936, mathematicians had demonstrated that the theory of general relativity implied that either an open or a closed universe was possible. The question would have to be settled by future observation.

One can determine the "size" of the universe in two ways: One can determine the average density of matter, or one can determine if the expansion is slowing down rapidly enough to eventually halt. Although the first method looks more straightforward, it is really the more difficult. In 1936, astronomers could see only matter which emitted light. Neither radio astronomy nor x-ray astronomy nor any of the other types of astronomy which depended on the detection of kinds of radiation other than light had been invented. In other words, if something was not luminous, it could not be detected. And, of course, no one knew how much matter the universe might contain that was not luminous.

The initial attempts to find out whether the universe was open or closed depended, therefore, on discovering how fast the expansion was slowing. But before that could be done, scientists had to discover what was wrong with Hubble's theory.

This feat was accomplished during the 1950s by two American astronomers, Allan Sandage and the German-born Walter Baade. During World War II, Baade had discovered that two different kinds of stars existed, which he called *Population I* and *Population II*. The Population II stars appeared to be much redder, and a spectroscopic examination of their light showed that they lacked the heavy metals present in stars of the Population I variety. Today we know that the stars of Population II are very old, while Population I stars were formed much more recently.

In 1944, Baade proposed that variable stars should also be divided into two types, and that two types of period–luminosity relations existed. He was not able to confirm this hypothesis, however, until he was able to make more accurate observations, using the new 200-inch telescope which was installed on Mount Palomar in 1947.

But in 1952, Baade announced that he had found the flaw in Hubble's work: There were two kinds of Cepheid variables, not one as Hubble assumed. Those of Population II were intrinsically brighter than those of Population I. When Hubble attempted to measure galactic distances, he confused the two, comparing Cepheids of one type in the Milky Way to those of the other type in other galaxies of the Local Group. The resulting errors in the distances of the nearest galaxies led to errors in the distances of all galaxies.

A few years later, Allan Sandage showed that Hubble had confused clouds of glowing hydrogen gas in distant galaxies with bright stars. By using these gas clouds as distance indicators, Hubble had introduced another systematic error.

The net result was that Hubble underestimated the size of the universe. Furthermore, the expansion was considerably slower than he had believed. Hubble's conclusion that the universe was expanding had been perfectly correct, but he had been wrong about intergalactic distances, the rate of expansion, and consequently the universe's age. New calculations gave a Hubble time of more than 10 billion years (later corrected to 18 billion). At last astronomers had a figure that was consistent with the age of terrestrial rocks.

Not content with just correcting Hubble's results, Sandage studied redshifts for decades, refining his methods at every step. He believed that precise and accurate observations would eventually enable him to determine the age of the universe and to discover whether it was open or closed.

In an article published in *Physics Today* in 1970, Sandage described his life's work as "a search for two numbers." One of these numbers, called the *Hubble constant*, is a measure of the rate of expansion. The other, called the *deceleration parameter*, is related to the rate at which gravity is causing the expansion to slow down.

The deceleration parameter is defined in such a way that it is less than one-half for an open universe, and greater than one-half if the universe is closed. The deceleration is greater in a closed universe because there is more matter and hence more gravitational braking. The figure one-half has no particular significance—it does not indicate that one quantity is one half of another. If the definition of the deceleration parameter were slightly different, the critical figure could be one-tenth, or five, or a hundred.

By 1956, Sandage had succeeded in measuring the distances of galaxies that were a billion light years away. At such distances, the individual stars in galaxies cannot be distinguished. Hence bright stars can no longer be used as distance indicators. Sandage was able to solve the problem by using the galaxies themselves.

Galaxies come in clusters, and many clusters contain

typical members of known brightness. At short distances, Sandage did what Hubble did: He calculated distance by measuring the apparent luminosity of bright stars. At large distances, he did something similar: He calculated the distances of clusters by measuring the apparent luminosity of bright galaxies.

When astronomers look far out into space, they are also looking far into the past. A galaxy a billion light years away is seen by light that has been traveling through space for a billion years. This is simply a consequence of the definition of the term "light year": the distance that light will travel in one year.

When Sandage looked at the nearer galaxies, he was seeing the universe as it had been in the recent past. When he examined distant clusters, he was viewing the cosmos as it had been a billion or more years ago. Since the universe was expanding more rapidly then, it was possible—in principle—to compare the two observations and to compute how much deceleration there had been.

In practice, it wasn't all that easy. The measurements were difficult, and the uncertainties were large. The step-by-step procedure used to measure distances—Cepheids first, then bright stars, then bright galaxies—made the task difficult, and variations in brightness of the various distance indicators made it even worse.

Sandage worked on the problem for years. By the mid-1950s he had a sufficient amount of data, or so it seemed. Measurements indicated that the deceleration parameter was approximately equal to one, twice the critical value of one-half. There were large uncertainties, but it seemed likely that the universe was closed.

In order to obtain a result, Sandage had to make certain

assumptions. One was that the brightness of galaxies remained relatively constant in time. There was no way that
Sandage could prove that a galaxy should be about as
bright now as it was a billion years ago, but the idea
seemed reasonable.

In the early 1970s, Sandage revised his figure to 1.2 for
the deceleration parameter. In 1978, Sandage and his collaborators, James Kristian of Hale Observatories and James
A. Westphal of the California Institute of Technology, revised this figure upward to 1.6. The three astronomers
pointed out, however, that even though 1.6 was more than
three times the critical value, one could not infer that the
universe was closed. If the brightness of galaxies changed
as time passed, zero might be a more accurate figure.

Although Sandage was one of the early proponents of
the closed universe, in recent years he has tended to favor
the conclusion that it is open. He points out that there are
other data (which I will describe in subsequent chapters)
which favor an open universe. He believes that more
weight must be given to these results than to his own deter-
mination of the deceleration parameter.

Recently the interpretation of some of Sandage's results
became even more uncertain when a group of astronomers
challenged the accepted value of the Hubble constant, the
number that measures the rate at which the universe is expanding. If they are correct, not only is the universe expanding more rapidly than astronomers have believed, it is
also somewhat younger than they have estimated. The new
results indicate that the cosmos is not 13 to 18 billion years
old, but only 7 to 10 billion.

In 1979, Marc Aaronson of Steward Observatory in
Tucson, Arizona; Jeremy Mould of the Kitt Peak National

Observatory; and John Huchra of the Harvard-Smithsonian Center for Astrophysics developed a new method for measuring intergalactic distances which gave different results from those obtained by Sandage and other astronomers. According to Aaronson, Mould, and Huchra, the standard measurements of the Hubble constant are thrown off by the fact that the expansion of the universe in the vicinity of the Milky Way is different from the expansion of the universe as a whole. The three astronomers say that we seem to be in a pocket of space that is expanding more slowly.

The anomaly, they contend, is caused by a massive cluster of galaxies in the constellation Virgo. They believe that this cluster is so massive that it retards the expansion of the universe in our region of space. The Milky Way is being drawn toward the cluster at a velocity of 400 to 500 kilometers per second; we are destined to collide with it in about 60 billion years.

The distances to galaxies have traditionally been determined in a stepwise manner. First, one measures the distances to galaxies in the Local Group, then the distances to nearby clusters, and, finally, the distance to clusters that are farther away. Now, one of these steps consists of the measurement of the distance to the Virgo cluster. If the anomaly in the expansion in our neighborhood throws these measurements off, other results will be in error also. This will lead, in turn, to errors in the values of the Hubble constant and the age of the universe.

It is as though a carpenter used an object one inch long to calibrate the length of one-foot rulers, and then used the rulers he had made to fashion yardsticks and tape measures. If he allows systematic errors to creep in at any step, then the biggest measuring devices—the tape measures—will not be accurate, and he will have problems when he tries to build a house.

Aaronson, Mould, and Huchra say that they have avoided such difficulties by developing a method that allows them to measure distant galaxies without going through the different steps. Their method is based on the fact that all galaxies rotate. The rate at which galaxies spin is related to their mass. Aaronson, Mould, and Huchra use measurements of radio waves and of infrared light to determine how fast galaxies are spinning. This tells them the mass. Once they know the mass, they can calculate the luminosity. Finally, a comparison of a galaxy's intrinsic luminosity with its apparent brightness tells its distance.

Aaronson, Mould, and Huchra claim that their method gives distances that are more accurate than those obtained previously. Other astronomers, at least those who use the more traditional methods, disagree.

At the moment, the new results are the subject of considerable controversy. A number of objections have been made to the Aaronson-Mould-Huchra theory. One of the most telling is the charge that the theory gives a figure for the age of the universe which is not consistent with that obtained from radioactive dating of the elements.

Most people are aware that radioactive dating can be used to establish the age of an object such as a fossil or a sample of rock. It is less well known that similar methods can be used to determine the age of the radioactive elements themselves. In other words, it is possible not only to find out how old a rock is by making measurements on the uranium in it, it is possible to find out how old the uranium is.

THE FATE OF THE UNIVERSE

It is believed that the universe originally consisted of nothing but hydrogen and helium. According to currently accepted theories, the heavy radioactive elements were created in supernova explosions. Some of this material is still drifting through interstellar space, but some of it has been incorporated into second-generation stars and the planets that formed around them.

If we find that a certain chunk of uranium is, say, 10 billion years old, we can conclude that the universe must be at least that age, and probably a few billion years older. There certainly cannot have been supernova explosions before there was a universe.

Results obtained from radioactive dating seem to indicate that the universe is at least 11 billion year old. This is consistent with the accepted figure of 18 billion years, but not with the 7 to 10 billion that Aaronson, Mould, and Huchra give. However, the uncertainties are just great enough that the controversy has not yet been resolved. In particular, the figures of 10 billion and 11 billion seem tantalizingly close. One cannot help but wonder whether future observations will lead to a small modification in one or the other, eliminating the inconsistency. For this to happen, we would only have to discover that radioactive elements could be a billion years younger than scientists currently think they are. Alternatively, a modification in the Aaronson-Mould-Huchra theory could conceivably give rise to a figure for the age of the universe that was a billion years greater.

If the Aaronson-Mould-Huchra theory turns out to be correct, it would have important implications for the question of an open or closed universe. If the universe is expanding more rapidly than astronomers currently believe, it is less likely that gravitation will ever bring the expansion to a halt. A rigorous calculation indicates that if the theory is correct, the value of the deceleration parameter is 0.35. Since this is less than the critical value of one-half, or 0.5, the implication is that the universe is open.

Thus we have a situation in which Sandage gives a value of 1.6 for the deceleration parameter, but admits that he does not have much confidence in it. Meanwhile, a theory that has not yet gained acceptance gives a value of 0.35. Both of these results depend on the debatable assumption that the brightness of galaxies remains constant over billions of years.

One can only conclude that if scientists are to find out whether the universe is open or closed, they will not achieve this success through measurements of the rate at which the expansion is decelerating. If the question is to be answered, more reliable methods must be found.

Fortunately, such methods are being developed. But before I discuss them, it will be necessary to say a few things about the way in which the universe began.

CHAPTER 3

The Big Bang

I F the universe has been expanding for billions of years, then there must have been a time when it was very condensed. At that time, the density of matter, the temperature, and the pressure were at extremely high levels. When the universe began, in other words, it was very dense, very hot, and already in a state of rapid expansion.

This explosion, which took place 13 to 18 billion years ago (or 7 to 10 billion years ago, depending upon which theory you believe), is called the *big bang*. It is not an inappropriate term, for the universe began as an enormous fireball. Since that time, it has been exploding outward. One should not think that the fireball burst outward into a previously empty space; space itself was created in the initial explosion. It was no more possible to get "outside" of the universe then than it is now.

It would be interesting to find out exactly how dense and how hot the universe was when it began. Unfortunately, there seems to be no way to do that. If one attempts to use the equations of general relativity to calculate just how compressed everything was, one obtains the result that it was initially infinite, that the entire universe was compressed into a mathematical point called the *singularity*.

Such a state of affairs is difficult enough to visualize in the case of a closed universe. But if the universe is open, things are even worse. In the latter case, one must imagine that infinite space and an infinite quantity of matter were infinitely compressed, that an infinite number of atoms were squeezed into a point.

Infinite quantities can be troublesome. Throughout the ages, the concept of infinity has led to one philosophical and mathematical paradox after another. And when physicists encounter infinite quantities in their calculations, they generally conclude that there must be limits to the applicability of the theory they are using. They do not normally take this to be an indication that the theory is wrong, because no theory is universally applicable. On the contrary, theories are approximations that sometimes break down when conditions become extreme. When this happens, physicists continue to use the theory under ordinary circumstances, and look for a new one that will work when conditions are extraordinary.

Einstein realized that there were limits to general relativity. In his 1950 book *The Meaning of Relativity*, he commented:

> The theory is based on a separation of the concepts of the gravitational field and matter. While this may be a valid approxima-

tion for weak fields, it may presumably be quite inadequate for very high densities of matter. One may not therefore assume the validity of the equations for very high densities and it is just possible that in a unified theory there would be no such singularity.

In other words, in a better theory, the prediction that the universe began in a condition of infinite density might disappear. Today scientists have some ideas about what such a better theory might be like. They even have a name for it; they refer to it as a theory of *quantum gravity*, because they suspect that such a theory would somehow combine general relativity and *quantum mechanics*. Unfortunately, no one knows how such a merger could be accomplished.

Quantum mechanics deals with the behavior of matter of the microscopic level. It is the theory on which virtually all of modern physics is based. Originally developed in the mid-1920s to explain the emission of light by atoms, it is now applied to such diverse areas as the structure of the atomic nucleus, the theory of chemical bonds, the behavior of transistors, and superconductivity in metals. Quantum mechanics is also the foundation on which theories of subnuclear particles are built. This list could be extended indefinitely, for quantum mechanics is applied to practically everything.

Quantum mechanics and the two theories of relativity *are* modern physics. Quantum mechanics and special relativity were combined decades ago, but combining quantum mechanics and general relativity is so difficult that physicists often feel they don't know where to begin.

As a result, no one can say exactly what was happening

during the early stages of the big bang explosion. Scientists cannot extrapolate all the way back; at the present time there is a theoretical barrier beyond which they cannot pass. In particular, they cannot say whether the universe was created in the big bang, or whether it previously existed in another form. They can't even say whether time existed "before" the big bang; for all they know, time may have been created in the explosion. In spite of this difficulty, the big bang theory has become almost universally accepted during the last decade. Competing theories do not seem to be compatible with scientific observations.

The big bang theory was developed over time, and has generated much controversy in the course of its history.

In 1922 a Russian mathematician named Alexander Alexandrovitch Friedmann found a mistake in Einstein's paper on the static universe. After he had corrected the error, Friedmann looked very closely at the new equations that he had obtained. He realized that Einstein's "static universe" really wasn't static at all. Only a very small perturbation would be required to nudge it in one direction or the other. Give it a little push, and it would quickly become either an expanding or a contracting universe.

Since Einstein's results were unsatisfactory, Friedmann set to work to find out just what kinds of universes the theory of general relativity did allow. In particular, he wanted to find out what the universe would be like if there was no cosmological constant.

In the end Friedmann succeeded where Einstein had failed. In 1922 he published in a German physics journal a paper entitled "On the Curvature of Space." The paper was only ten pages long, but in those ten pages Friedmann gave, for the first time, a correct mathematical description of the universe.

It was Friedmann who discovered that there are two kinds of universes, those that we call "open" and "closed." He showed that the average mass density determined the type of a universe. Finally, he showed that the universe must be either expanding or contracting. But his work was largely ignored.

Shortly after Friedmann's death in 1925, the abbé Georges Lemaître, a professor of relativity and the history of science at the University of Louvain, began to consider the same problems that Friedmann had worked on. Since he had never heard of the latter's work, Lemaître had to carry out the same calculations all over again. He finally achieved success, and in 1927 he published a paper on the subject. Like Friedmann's, it too was ignored.

In 1933, Lemaître proposed that the universe originally consisted of a "primeval atom" which contained all of the matter that was later to become the universe. At a given point in time, this primeval atom disintegrated, sending matter flying off in every direction. The creation of the universe, in other words, was envisioned as a process that resembled nuclear fission. His primeval atom split apart in a manner similar to the way in which uranium or plutonium atoms fission in an atomic bomb explosion.

Lemaître's theory was correct in principle, and wrong in all the details. Today we know that the universe never existed as the kind of solid ball that Lemaître conceived, and that it did not fly apart in the manner he thought. Nevertheless, Lemaître has to be acknowledged as the father of the big bang theory of the origin of the universe. It was he who first drew the conclusion that if the universe was expanding, then it must have once existed in a very condensed state.

About a year after Lemaître published his theory, one of Friedmann's former students, a physicist named George Gamow, emigrated from the Soviet Union to the United States. If Lemaître was the father of the big bang theory, Gamow was the scientist who changed it into something resembling its modern form. It was also Gamow who invented the term "big bang."

On April Fool's Day in 1948, a paper appeared in the prestigious scientific journal *Physical Review.* The paper had apparently been authored by three scientists, named Alpher, Bethe, and Gamow. In reality, the paper had been written by Gamow and his student Ralph Alpher. Cornell University physicist Hans Bethe didn't even know about it until he saw it in print. It seems that Gamow, who was something of a practical joker, had added Bethe's name without his knowledge so that the paper would have an authorship that was a pun on the first three letters of the Greek alphabet.

According to the Alpha-Beta-Gamma theory, as it came to be known, the explosion that took place at the creation of the universe resembled that of a hydrogen, rather than an atomic, bomb. At the time, of course, few people knew what a hydrogen bomb was; the first one was not to be detonated until 1952. However, in 1948 it was known that nuclear fusion was possible. Scientists had realized for some time that the lighter atomic nuclei, such as those of hydrogen, could be fused together to make heavier nuclei, like helium, releasing energy in the process.

In the beginning, according to Gamow, the universe was

made up of neutrons. The neutrons were packed together, not in a cold primeval atom, but in a fireball whose temperature was well over a billion degrees Celsius. The universe was initially so hot that there existed far more energy than matter.* The moment it was formed, this fireball began simultaneously to expand and to cool.

One can't make a universe out of energy and neutrons alone. Atoms have three main constituents: neutrons, protons, and electrons. The neutrons and the protons are found in the atomic nucleus; the electrons circle around them. Fortunately there is a process called *beta decay*. If left to itself, a free neutron will, after about fifteen minutes, transform itself into a proton, an electron, and a neutrino. The term "free neutron" is important: Neutrons which are bound in atomic nuclei are much less likely to undergo this transformation.

In Gamow's universe, once about half of the original neutrons had undergone this beta decay process, all the materials needed to make ordinary matter would be present. Gamow's theory was intended not only to explain the origin of the universe, but also *nucleosynthesis*, or the formation of the elements.

Even though Gamow's theory seemed to work a little better than Lemaître's did, scientists soon began to suspect that there was something wrong with it also. Detailed calculations showed that although hydrogen and helium could easily be formed in Gamow's universe, conditions would never be quite right to make the heavier elements. If his theory was correct, then there should be no carbon, ni-

---

* Einstein's formula $E = mc^2$ asserts that matter and energy are equivalent; hence the two can be compared directly. In this equation "E" is the amount of energy, "m" is the equivalent mass, and "c" is the speed of light.

trogen, silicon, copper, iron, or any of the other elements with which we are so familiar. For that matter, there would be no George Gamow. Something had to be wrong with Gamow's conception.

A possible solution was provided when British astronomer Fred Hoyle published a paper in which he hypothesized that the heavy elements had been "cooked" in the interiors of stars and then spread through space by supernova explosions. Today we know that Hoyle's theory is entirely correct. With the exception of small amounts of light substances like deuterium (heavy hydrogen) and lithium, only hydrogen and helium were made in the big bang. Everything else is a product of the nuclear reactions which take place in the interiors of stars.

But there was one respect in which Gamow's theory turned out to be wrong. By the early 1950s it had become apparent that the universe was not originally a sea of neutrons. Temperatures in the primeval fireball were so high that neutrons and protons must have been created in roughly equal amounts.

According to Einstein's formula $E = mc^2$, matter and energy are interchangeable. If enough energy is present, matter can be created out of empty space (this is just the opposite of what happens in the detonation of a nuclear bomb, where energy is created out of matter).

The energy density of the big bang fireball was enormous. A hundredth of a second after the beginning of the universe, the temperature was approximately 100 billion degrees Celsius (Gamow's billion degrees was an underestimate). As the universe expanded, it cooled rapidly. But for a short time, the energy in the universe was so concentrated that neutrons, protons, and subatomic particles of every kind were created. Gamma rays disappeared and matter was created. Much of this matter was quickly transformed into energy again, but enough neutrons, protons, and electrons remained to form the matter that fills the universe today.

After the universe had been expanding for several minutes, the temperature dropped to a billion degrees, cool enough for the formation of helium nuclei. But there were not yet any atoms.

Atoms form when electrons are able to attach themselves to nuclei. For example, a helium nucleus consists of two protons and two neutrons. A helium atom consists of two electrons which circle the nucleus. In the early universe, the temperature and energy density were so high that electrons were knocked out of their orbits the instant the atoms formed.

It was not until 700,000 years had passed that the universe cooled enough—to about 3,000 degrees—to allow the existence of stable atoms. At this point the universe, which had previously been opaque, suddenly became transparent. Once the electrons were bound in atoms, light could pass through it freely. Before that time, light had bounced off the electrons which filled all space.

More time passed. After hundreds of millions of years— or possibly billions; no one knows exactly—the hydrogen and helium condensed into galaxies, and then into stars. The more massive stars burned out quickly and exploded as supernovae, permeating space with the heavy elements that had been cooked in their cores. Second-generation stars formed. Planets were created. And before long there was life.

The big bang theory deals with events that we will never

be able to observe, as do many other scientific theories. No one has ever seen a neutron or an electron, for example, and no one ever will; subatomic particles are far too small to be observed in microscopes. Even if we could manage to make optical instruments capable of the necessary magnification, we still would not be able to see them. The laws of quantum mechanics place limits on what one can hope to observe. Yet physicists speak of subatomic particles as though they were visible. This is because, though we cannot see electrons, we can see the things that they do. A television picture is created when a beam of electrons strikes the phosphor inside the picture tube. An electric current is created when a stream of electrons passes through a wire. And, finally, individual electrons can be counted by particle detectors of various kinds. When the electron hits the detector, things happen (exactly what kinds of things depends upon the type of detector). The resulting signal can be amplified, giving us evidence that an electron has arrived.

The theory that matter is made up of protons and neutrons and electrons provides us with excellent explanations of things that we encounter every day of our lives, as well as of the events that are made to take place in scientific laboratories. Because the theory does such a good job of explaining the things that we can see, scientists also believe what it says about the particles that are too small to be seen.

An analogous situation exists with theories about the universe. If a theory successfully explains why the universe is the way it is, we generally believe what it says about the things that happened in the past, even if they took place billions of years ago.

THE FATE OF THE UNIVERSE

The big bang theory explains why the universe should be expanding. It tells us the origin of the matter that makes up stars and galaxies, and why the universe is mostly hydrogen and helium. And, finally, it makes two very important predictions which have been confirmed by scientific observations.

The first of these has to do with the amount of helium in the universe. Helium is observed on the surface of the sun, in the stars that make up our galaxy, in other galaxies, and in interstellar space. It is also found in the cosmic rays (which are really not "rays," but rapidly moving particles) that continually bombard the earth. Most astronomical objects whose chemical composition is known are between 23 and 27 percent helium, and it is estimated that the universe is about 25 percent helium by weight. Most of the rest is hydrogen, and there are small amounts of other elements.

The nuclear fusion reactions which power the stars convert hydrogen into helium, but do not make enough to account for the amounts observed. If helium were made only in stars, the universe would be only a few percent helium, not 25 percent. Therefore, most of the helium which exists must have been made some other way.

The big bang theory says that between 20 and 30 percent of the matter in the universe should have been converted into helium during the first few minutes of the expansion. As we will see later, the amount depends on exactly how fast the universe was expanding initially. For the moment, I want only to point out that the observed amounts of helium provide us with an excellent confirmation of the big bang theory.

The helium was formed in the following manner: First,

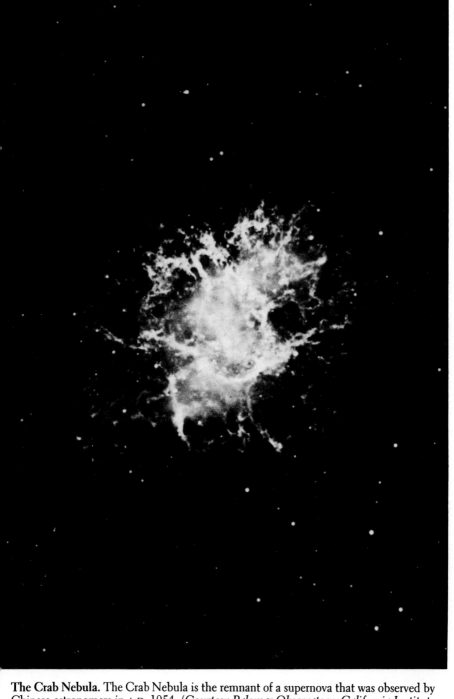

The Crab Nebula. The Crab Nebula is the remnant of a supernova that was observed by Chinese astronomers in A.D. 1054. (*Courtesy Palomar Observatory, California Institute of Technology.*)

Cluster of Galaxies in Coma Berenices. In 1933, American astronomer Fritz Zwicky pointed out that although the cluster was apparently held together by gravity, the galaxies did not have enough mass to provide the necessary attractive force. This "missing mass" problem has been solved only in the last few years. It has been discovered that the greater part of the mass in the cluster exists, not in the galaxies, but in massive dark haloes that surround them. (*Courtesy Kitt Peak National Observatory.*)

The First Known Gravitational Lens. These two photographs, printed in negative form for greater clarity, cover the same region of the sky. The second, which was taken at a longer exposure, shows objects too faint to be seen in the first.

The first photograph shows the double quasar 0957 + 561. The two images, marked "A" and "B," have the same red shift and identical spectra (they emit light at the same wavelengths). The second photograph shows a cluster of galaxies between the double quasar and the earth. The five brightest galaxies are marked G1 through G5. The B image is "buried" under the image of G1. From analysis of this picture and other data it has been determined that the double quasar is really two images of a single quasar. The splitting is caused by the gravitational fields created by G1.

This was the first known example of the gravitational-lens effect, which was predicted by Einstein in 1936. The double quasar was discovered in 1979.

These pictures were taken by Jerome Kristian, James A. Westphal and Peter Young, using the 200-inch telescope at Palomar Observatory. They were made with a CCD (charge coupled device) solid-state television detector developed by Texas Instruments for use on the Space Telescope, which is due to be put into orbit in 1984. (*Courtesy Dr. Jerome Kristian, Mt. Wilson and Las Campanas Observatories, Carnegie Institution of Washington.*)

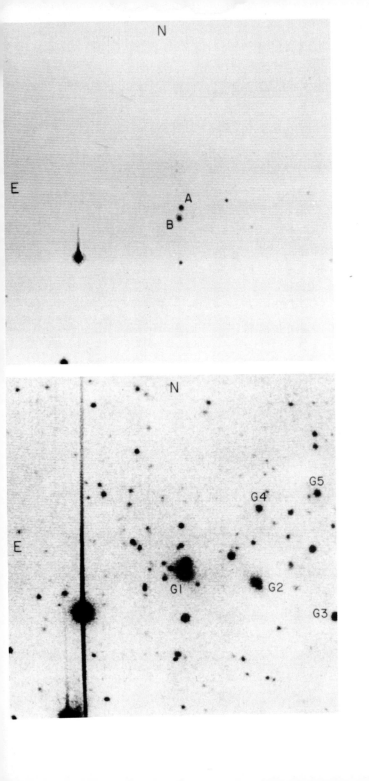

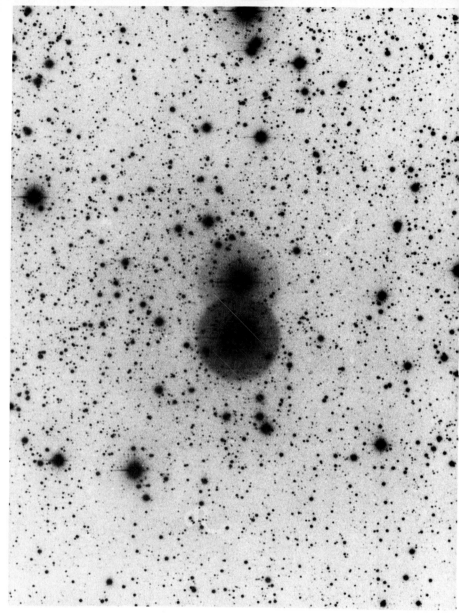

Cygnus X-1. The lower star in the center of this photograph (printed in negative form) is HDE 226868, the blue giant whose invisible companion is believed to be a black hole. The photograph was taken through the 200-inch telescope at Palomar Observatory by Dr. Jerome Kristian. (*Courtesy Dr. Jerome Kristian, Mt. Wilson and Las Campanas Observatories, Carnegie Institution of Washington.*)

neutrons collided with protons to form deuterium, or heavy hydrogen (a deuterium nucleus is made up of one neutron and one proton). When deuterium collided with another neutron, tritium, another isotope of hydrogen, was formed (tritium is composed of two neutrons and one proton). Finally, the tritium collided with a second proton to make helium. In describing this process, I have ignored the electrons. At this point, it was still too hot for the electrons to be captured by the nuclei.

The universe was now made up of helium and free neutrons and protons. But, as I have pointed out previously, a free neutron will decay in about fifteen minutes, and a proton, an electron, and a neutrino will appear in its place. In the early universe, temperatures were so high that this and similar processes took place even more rapidly. After a few minutes, the universe consisted mostly of helium nuclei, and protons and electrons, which eventually combined to make hydrogen gas. An atom of hydrogen is nothing more than a proton and an electron held together by gravitational attraction.

If measuring the amount of helium present in the universe were the only way to confirm the big bang theory, then the theory would not be widely accepted. It would still be possible to invent alternative explanations for the presence of the helium. But there is another confirmation of the big bang theory, one that is so convincing that astronomers have given up as a waste of time attempts to invent alternative explanations for the helium.

In 1964 two Bell Telephone Laboratories scientists, Arno Penzias and Robert Wilson, made one of the most significant discoveries in modern astronomy: They observed the light emitted by the primeval fireball.

What Penzias and Wilson observed was not a brilliant flash of light, but simply a background of radio waves. The light from the fireball had been traveling through space for many billions of years. During that time it had undergone enormous redshifts, losing energy until it could be observed only in the radio part of the spectrum.

These radio waves come from all directions in the sky. This is not surprising, because the big bang happened everywhere, filling the whole universe. If the universe is finite, the big bang occupied all of a finite space. And if the universe is infinite, then the big bang filled infinite space.

If the radio wave background that Penzias and Wilson detected did not have approximately the same intensity in all directions, then it could not be a remnant of the primeval fireball. If it came from any particular direction, or was more intense in one direction, then we would be forced to conclude that it had nothing to do with the big bang, that the radio waves had been produced by something that happened in a particular part of the universe.

The evenness of the radio wave background is only one confirmation of its origin. It also varies in intensity with wavelength in precisely the way that the big bang theory says that it should. Since 1964, numerous measurements have confirmed that the background is, indeed, a relic of the light from the fireball.

This confirmation of the big bang theory is even more compelling than that provided by measurements of the helium content of the universe, because in the case of the helium it is possible to invent alternative explanations. Here it is not. Some have been put forward, but none have withstood scientific inquiry.

Astronomers refer to this background as the *cosmic mi-*

crowave background radiation. The name is long, but straightforward. "Microwave radiation" is a name for radio waves with wavelengths less than one meter. "Background" means it is everywhere, and "cosmic" is a reference to its origin.

The existence of a microwave background had been predicted in a paper published in 1949 by Gamow's collaborators Ralph Alpher and Robert Herman. Alpher and Herman had calculated that the background should be made up of *blackbody radiation* at a temperature of 5° K.

This terminology requires a few words of explanation. "K" stands for *Kelvin,* a temperature scale that is often used by scientists. Degrees in the Kelvin and the Celsius temperature scales are really the same; the difference between the two scales is the location of the zero point. Zero on the Kelvin scale is known as absolute zero, and is the theoretically lowest possible temperature, while zero on the Celsius scale is the freezing point of water. Since absolute zero is −273° C, 0° C equals 273° K, 100° C equals 373° K, and so on.

"Blackbody radiation" is radiation that depends on temperature. Any object that is not at a temperature of absolute zero will emit radiation. A hot piece of metal will become incandescent and give off visible light; one that is only warm will emit infrared radiation; and one that is cold will produce microwaves.

A blackbody is a hypothetical body that would absorb all radiation without reflecting any of it back. In nature there is no such thing as a perfectly black body. Every real object reflects some of the light that falls on it. However, it is possible to construct laboratory apparatus that emits light in exactly the way that a real blackbody would. If a tiny hole is punched in a closed box, the hole will be almost perfectly black. Any radiation that enters it will bounce around inside the box until it is absorbed. And if the box is heated, it will emit radiation which reproduces the theoretical blackbody spectrum perfectly.

When Penzias and Wilson set up their experiment, they were not looking for blackbody radiation from the big bang. They were not even trying to perform astronomical observations. They were simply trying to eliminate the static in an antenna that was used for communication with satellites.

Penzias and Wilson were not aware that Alpher and Herman had predicted a microwave background. The same familiar story had repeated itself; the paper that predicted background radiation had been forgotten.

A group of scientists at Princeton University had rediscovered Alpher and Herman's results, and were preparing to set up an experiment to look for the background. But by this time Penzias and Wilson were already detecting the radiation, though they did not know what it was.

At first, Penzias and Wilson were certain that it had to have a source somewhere in the antenna, so they took the antenna apart and put it back together again. They cleaned out a "white dielectric material" that had been deposited by some pigeons. They masked rivets with aluminum tape. In spite of everything, an annoying hiss remained. Penzias and Wilson knew that this static could not be caused by anything in their vicinity, for it was the same by day and by night, and it did not change when the antenna was pointed in a different direction. Neither, as far as they could tell, did it come from space. The hiss never varied in intensity as the stars rose, passed across the sky, and sank over the horizon. Since the two scientists did not know about the prediction of a background, they never suspected that radio waves could have an extraterrestrial source and yet be absolutely unvarying.

In December of 1964, Penzias happened to mention the problem to astronomer Bernard Burke. Burke had heard about a talk given by P. J. E. Peebles, a member of that group of Princeton physicists who had rediscovered Alpher and Herman's results. In the talk, Peebles had mentioned that there ought to be a background of microwave radiation left over from the big bang.

Penzias soon learned that the annoying static was not a result of some defect in the antenna, or of loose rivets, or even of the pigeon droppings. Penzias and Wilson had been hearing the universe itself.

Penzias and Wilson measured the microwave background at only one frequency, so they could not be absolutely sure that they had observed blackbody radiation. But subsequent measurements by various groups of scientists confirmed the fact that the background varied in intensity at different frequencies in exactly the manner that blackbody radiation should.

The background that Penzias and Wilson discovered resembles that which would be emitted by a blackbody at a temperature of about three degrees above absolute zero. Hence it is sometimes referred to as the 3K *radiation background*. But one should not conclude that 3° K is the average temperature of matter in the universe. On the contrary, it is a radiation temperature. The microwave background, after all, was not recently emitted by bodies with temperatures of $3°$ K. It has been traveling around the universe for 18 billion years. During that time, it has interacted with matter very little or not at all.

Measurements of the helium content of the universe and the discovery of the cosmic microwave background confirmed the fact that there had been a big bang, but it was not known whether the big bang had taken place in an open or a closed universe. However, until the mid-1970s, astronomers tended to think that it was closed.

It is not easy to say why they favored this conclusion. It is true that both Einstein and Hubble had believed in a finite universe. But, as we have seen, there were flaws in their work. Sandage's measurements of the deceleration parameter did seem to argue for a closed universe, but these results were anything but conclusive.

Perhaps the tendency to favor a finite universe was partly the result of a philosophical prejudice. A closed universe seems somehow to be tidier than an open one. One does not have quite as many problems as with the infinite. And one does not have to contemplate a universe that will eventually die.

In an open universe, life does not go on forever. As the universe expands, the stars burn out and galaxies grow dimmer. For a while, new stars are created in ever decreasing numbers. But eventually the supply of interstellar gas from which stars are created runs out, and star formation stops. The last stars are able to continue shining for billions of years. But in the end, they too burn out and the universe becomes a cold, dark waste.

A closed universe, on the other hand, could very well go on forever. It will eventually collapse, to be sure. But who is to say that it will not re-explode in a new big bang? If the collapsing universe could somehow "bounce" and enter into a new phase of expansion, life might be created anew after every big bang. If this did happen, it would be reasonable to expect that intelligent beings would exist in every cycle. After all, there is at least one intelligent species inhabiting this cycle, and there is no reason to think that conditions would be very different the next time around.

Somehow we humans dislike contemplating the idea that intelligent life is something destined to exist for only a brief span in the history of the cosmos. This fact was probably partly responsible for Einstein's and for Hubble's preference for a closed universe.

During the 1970s, evidence began to accumulate which appeared to contradict Einstein's and Hubble's and Sandage's hypothesis that the universe must be closed. The turning point was a paper that four astronomers—J. Richard Gott III and James E. Gunn of the California Institute of Technology, and David N. Schramm and Beatrice M. Tinsley of the University of Texas—published in *The Astrophysical Journal* in 1974. In this paper, Gott, Gunn, Schramm, and Tinsley analyzed all of the available evidence, and concluded that the universe was very probably open.

One of their arguments was based on the amount of matter in the universe. According to the four astronomers, the best measurements of the average mass density gave a figure that was far too low for a closed universe. There was just not enough matter to bring the expansion to a halt.

Another argument, which the authors believed to be even more convincing, had to do with primordial deuterium. Measurements indicate that the ratio of deuterium to ordinary hydrogen in the universe is about 20 or 30 parts per million. Like helium, deuterium is present practically everywhere that astronomers look. The presence of deuterium invites certain conclusions.

It seems clear that the deuterium had to have a primordial origin; it had to have been made in the big bang. Unlike helium, and unlike the heavier elements, deuterium cannot be created by the nuclear reactions that take place in stars. Although the deuterium nucleus is stable under the conditions that prevail on the surface of the earth, it is not stable at the temperatures that exist in stellar interiors. If any deuterium is formed in stars, stellar temperatures cause it to break apart into its constituent neutron and proton almost at once.

Since there is no other conceivable way that the deuterium could have been made, the deuterium that is seen today must be a remnant of the big bang. The big bang theory tells us, in fact, that the creation of deuterium was a step in the process by which helium was made. The theory even tells us why so little exists: Most of it was converted into helium, or broken apart by collisions.

The big bang theory also tells us that the *amount* of deuterium has important implications. Calculations show that there is a relationship between the density of matter in the early universe and the amount of deuterium that remained after the nuclear reactions that took place in the primeval fireball were over. The denser the early universe was, the more often the deuterium would have undergone collisions that would have either broken it apart or converted it into helium.

Since an open universe is less dense than a closed one, more of the deuterium emerges from the fireball unscathed. If the universe is closed, then we should expect to observe only a fraction of the 20 to 30 parts per million that we do see. Unless there is some other way to explain how this deuterium managed to survive, we must conclude that the universe is open.

There is another way of phrasing this argument. Recall that it is not the matter density alone that determines whether the universe is open or closed, but rather the matter density for any given rate of expansion (when we talk of the present matter density, we are really comparing this density to the rate of expansion that is observed today).

This means that instead of saying that a closed universe is denser than an open one, we can say that a closed universe expands more slowly. The two statements are equivalent. This makes it possible to explain the presence of the deuterium; because an open universe expands more rapidly, the primordial matter remains packed together for a shorter time, and there is less opportunity for the deuterium to break up.

Since Gott, Gunn, Schramm, and Tinsley published their paper in 1974, theoretical work has continued, and it has been shown that the observed helium content also indicates an open universe.

Things are a little more complicated in this case. For one thing, some of the helium that is seen today was created in stars; it was not all formed in the fireball. For another, the helium content is less sensitive to the expansion rate than the deuterium content is. In other words, when we assume different possible rates of expansion, the percentage of deuterium changes much more rapidly than the amount of helium does.

However, we do know that there is approximately 25 percent helium in the universe today. Hence there must have been less than this when the universe emerged from the fireball. And if the universe was less than 25 percent helium by weight then it must be open.

The situation here is the opposite of that with respect to deuterium. An open universe contains more deuterium, but less helium. The longer matter remains packed together in the fireball, the more helium is made. Detailed calculations indicate that if the universe is closed, there should be more than 25 percent helium by weight.

Must we conclude, then, that the universe is open? Surprisingly, the answer to this question is no. Although it is not very likely that the deuterium and helium arguments will be refuted, it has become apparent in the last few years that they might contain loopholes, and that the questions concerning the nature of the universe have not been settled after all. I will discuss some of these loopholes in subsequent chapters.

CHAPTER 4

Missing Mass

T HERE is one important gap in the big bang theory: It does not tell us why there are stars and galaxies. In fact, the theory, in its pure form, seems to imply that galaxies should not exist. The initial explosion should have sent matter hurtling outward with such great velocity that it should never have had the chance to collect into the bright clumps of matter we see every time we look into the night sky.

There is only one way that galaxies could have been formed. Billions of years ago there must have been regions, hundreds of thousands of light years in diameter, where the density of the primordial hydrogen and helium gas was greater than it was elsewhere. Gradually, these clouds contracted under the influence of gravity. After hundreds of millions of years, they fragmented. The fragments contracted even more, forming regions of still higher density. As some of these were compressed still further by gravity, they became hot. One by one, the stars began to blink into existence.

It all sounds very natural, perhaps even inevitable. But there is something very paradoxical about this process. In order for galaxies to form, the universe must have been expanding and contracting at the same time. While the universe as a whole increased in volume, the opposite happened in the protogalactic clouds; they became smaller. If we are to explain how such a situation came about, the big bang theory must be modified.

Theoretical calculations show that the condensations that were to become galaxies must have existed very early in the history of the universe. They must either have existed from the beginning, or have formed sometime during the first several hundred thousand years. They could not have been created later because gravitational forces would no longer have been strong enough to overcome the general expansion.

The big bang, in other words, was lumpy. Furthermore, the lumps already existed during the early stages of the expansion, when conditions were so extreme that nothing resembled matter as we know it. The origin of galaxies can be traced back to clumps in a hot primordial soup that was made up of radiation and of subatomic particles that were constantly being created and destroyed.

No one knows where these clumps came from. The problem is not that we lack a theory, but that there are too many theories. At present, there seems to be no compelling reason why one should be favored over any of the others.

According to one theory, the early universe was very turbulent, and the galaxies were formed from eddies in the cosmological fluid. According to another, there was no single big bang, but rather a large number of "little bangs." These little bangs were supposedly responsible for the condensations that later became galaxies. According to yet another theory, magnetic fields produced the lumps.

There is also a theory which says that the lumps were melted-down galaxies from a prior universe. At one time this universe was presumably exploding like our own. After billions of years, it began to contract. Gravity accelerated the contraction until everything became so compressed that the collapse could not continue (the theory does not say why the collapse could not go on; it is assumed that collapse stops in a region where general relativity is no longer valid). At this point, this universe re-exploded in a new big bang. But the condensations that had been galaxies remained.

There are yet other theories. Although some seem more plausible than others, there is not yet any observational evidence that would allow scientists to decide among them. Until astronomers can devise some way to look far enough out into space—and hence far enough back in time—so that they can actually see galaxies being formed, they might not have any way to decide among the various theories.

This is unfortunate. If we knew how galaxies were made, we might be able to determine how much of the matter in the universe is condensed into galaxies, and how much exists in other forms. And if we could do this, we might be able to tell if the density of matter is high enough to close the universe.

We know how much matter there is in galaxies because

galaxies can be seen; they are made up of stars and glowing clouds of gas. It is the matter that exists outside of galaxies and that does not radiate enough energy to make itself visible to us that causes all the problems. We do not know how much of this dark, non-luminous matter there is, or what form it might take. The universe could be filled with clouds of gas, with dim extragalactic stars that are not bright enough to be seen, even with "snowballs" of frozen hydrogen. Until astronomers find ways to tell in what forms extragalactic matter exists, and to weigh it, no one will know for certain what the mass density of the universe is.

In the previous chapter I said that the paper by Gott, Gunn, Schramm, and Tinsley had stated that the observed matter density indicated that the universe was open. But those four authors could speak only of matter that could be seen. As astronomers are fond of pointing out, for all we know, intergalactic space could be filled with copies of *The Astrophysical Journal*; if they emitted no energy, there would be no way to detect them.

As a matter of fact, astronomers have known for approximately half a century that there is something out there that they cannot see. This fact was first pointed out in 1933 by California Institute of Technology astronomer Fritz Zwicky. Studying a large cluster of galaxies in the constellation Coma Berenices, Zwicky discovered that although the galaxies in the cluster were apparently bound by their mutual gravitational attraction, the amount of mass actually observed was only a fraction of that needed to hold the cluster together. Zwicky pointed out that there was a *missing mass* problem. Astronomers are only now beginning to solve it.

"Invisible mass" would undoubtedly be a better term.

After all, nothing is really "missing." The mass is very much there, but cannot be seen. However, "missing mass" has become the accepted terminology.

About 70 percent of all of the galaxies in the universe are bound in clusters, which range in size from small clusters such as the Local Group to *rich clusters* that have thousands of members. The Local Group is made up of two large spiral galaxies, the Milky Way and the great galaxy in Andromeda, and about twenty dwarf galaxies. A rich cluster is a very large group that has an especially high density of galaxies at its center.

Offhand, one might think that it would be possible to obtain an estimate of the mass of a galaxy by counting the stars in it. However, this is impossible. In the firt place, the stars are far too numerous. Even the smallest dwarf galaxies contain something like 100 million, while large spirals like the Milky Way have about 200 billion. And there are galaxies that are larger even than this. Giant elliptical galaxies can contain upward of a trillion stars.

Furthermore, the stars in galactic cores are so close together that even the most powerful telescopes cannot resolve them; one can see nothing but a brilliant blob of light. Meanwhile, the dimmer stars cannot be seen at all. It is hard enough to make out white dwarfs in our own galaxy; in other galaxies they are not visible.

Even if one could count all the stars, this still would not provide a reliable estimate of a galactic mass. Galaxies contain different amounts of interstellar gas and dust. Some appear to have undergone collisions which have stripped them of their gas; others have so much that it contributes significantly to their mass.

The method used to compute galactic masses is basically

the same method used to find the mass of our sun. The sun's mass can be determined by examining the motion of the planets in the solar system. If one knows the distance of a planet from the sun, and its velocity, then the mass of the sun can be computed.

It is the motion of the planets in their orbits that keeps them from falling into the sun. The closer a planet is to the sun, the faster it must move to balance the sun's gravitational pull. Thus Mercury will go around the sun in 88 days, while it takes Saturn 29 years, and Pluto 249 years. The time of revolution is related only to the distance from the sun and to the mass of the sun, not to the size of the planet. If a massive planet like Jupiter were as close to the sun as the earth is, its time of revolution would also be one year.

Galaxies are a bit more complicated than the solar system, but the situation is basically the same. The individual stars in a galaxy revolve around the galactic core. Their orbits are very large, and their periods of revolution are very long, but the principle is the same.

It is true that the mass of a galaxy is split up among many different bodies; the Milky Way contains 200 billion stars, while the sun has only 9 planets. But if we can measure the velocity at which the stars at the outer edge of a galaxy are revolving, and if we know the distance to the galactic center, we can compute the galactic mass.

This method allows astronomers to find not only the mass in the core itself, but all of the mass that lies within the orbits they examine. For example, if a star is 30,000 light years from the galactic center, then observations of its orbit allow astronomers to compute all of the mass out to that distance. They are able to do this because the star is orbiting a "smeared out" mass that includes the core and also all of the stars that are nearer the core than it is.

Obviously we cannot watch galaxies for millions of years in order to see how long it takes for any given star to make a complete circuit. Fortunately, one need only determine how fast a star is moving. For example, we know that our sun is traveling in its orbit at a velocity of about 250 kilometers per second. This fact makes it possible for astronomers to compute that its period of revolution must be 250 million years.

In order to determine how fast a star is moving, we measure its redshift (or blueshift). Once the proper corrections are made for the motion of our sun, and for the speed of recession of the galaxy in which the star is located, its velocity can easily be determined.

In other words, redshifts can be used not only to determine how fast the galaxies are moving away from us, but also how rapidly they are rotating. Since all of the stars in a galaxy revolve in the same direction, and since stars that are the same distance from the center move at the same speed, their motion has the appearance of a galactic rotation.

If we look at a galaxy edge on, one rim will appear to be moving away from us, while the other is approaching (to get an idea of what this looks like, just close one eye and look at a rotating phonograph record from the edge). Light from the stars on the rim that is moving away will be redshifted, while that emitted by stars on the other will be shifted toward the blue wavelengths. The red and blue shifts allow the velocity of rotation to be computed.

In practice, astronomers look at the light from clouds of

glowing hydrogen gas rather than at that emitted by stars when they want to measure rotation rates. However, these clouds rotate around the galactic center in the same manner that stars do, so this really doesn't change the principle involved.

Certain corrections have to be made when a galaxy is not seen exactly edge on (few are). But this is a calculation that involves nothing more than trigonometry. All in all, the method for measuring galactic mass is simple enough that any errors introduced when the measurements are made cannot be serious ones.

The method used to estimate the mass of clusters of galaxies differs only in that the motions of entire galaxies are measured. If one knows how fast the individual galaxies are moving (here we are talking about their motion around one another, not rotation), then it is not difficult to compute the mass that is needed to hold the cluster together.

In order to perform the calculation, one need only assume that the gravitational energy of the galaxies balances their energy of motion. One does not even have to make use of the equations of general relativity; Newton's law of gravitation is quite adequate here, and Newtonian physics is used to determine both galactic masses and the masses of clusters.

But if the masses of the individual galaxies within a cluster are added up, there is a serious discrepancy. The sum of the masses of the galaxies always turns out to be much less than the mass of the whole cluster. This is true of every cluster that astronomers have studied—the anomaly is not peculiar to the one that Zwicky studied in 1933. The amount of the missing mass may vary; nevertheless, it always exists.

Astronomers realized that there was something wrong either with their methods of measuring mass in galaxies, or with their determinations of the mass present in clusters. But the cause of the discrepancy remained a mystery until the 1970s, when new discoveries in radio and x-ray astronomy pointed the way to an answer.

Using methods similar to those which optical astronomers had applied to glowing gas clouds within galaxies, radio astronomers discovered that there was a significant amount of mass outside the visible parts of galaxies. The galaxies were surrounded by dark haloes. By the mid-1970s it had become apparent that there was more mass in the haloes than there was in the galaxies themselves.

Cool hydrogen gas in interstellar space emits radio waves at a wavelength of 21 centimeters. Now radio waves, like light, are redshifted or blueshifted when the material that emits them is moving relative to the observer. If the hydrogen gas is moving away from the earth, the wavelength will be redshifted to slightly more than 21 centimeters, and if the gas is approaching, the wavelength will be less.

The gravitational forces that cause stars to revolve around a galactic center will cause gas outside the galaxy to behave in the same manner. The only difference is that the gas moves in a larger orbit, and its time of revolution is consequently longer. But if one can measure the velocity of the gas, it is possible to compute the mass that lies inside its orbit.

It should not be inferred that the haloes are necessarily made up of hydrogen gas. It was obvious that some gas was present, but there was not enough of it to account for the masses that were computed. Something invisible was also orbiting the galaxies. What it was, the radio astronomers could not determine.

Scientists still do not know what it is. Neither do they know exactly how much mass the haloes contain, for it has not been determined how far out they extend. Naturally, numerous suggestions have been made as to what the invisible mass might be. One suggestion is that the haloes may be made up of small, very dim stars that give off too little light to be seen from the earth. According to another hypothesis, the haloes are composed of dark remnants of stars that burned out billions of years ago. One of the most intriguing possibilities is that the haloes might be made of subatomic particles called neutrinos. This hypothesis will be discussed in detail in Chapter 6.

While the radio astronomers were finding that some of the missing mass was located in galactic haloes, the x-ray astronomers were discovering that some of it existed in another, entirely different form. In 1970, satellite observations detected hot hydrogen gas in the supposedly empty spaces between galaxies in clusters. Measurements indicated that this gas had a temperature that was somewhere between 10 million and 100 million degrees Celsius (or Kelvin; when one talks of million-degree temperatures the 273-degree difference between the two scales becomes insignificant). Since gas at such high temperatures emits radiation only in the x-ray region, optical and radio observations had never been able to detect it.

Temperature is related to the average energy of motion of the molecules in a substance. When an object is heated, the molecules in it move more rapidly, and when it is cooled, they slow down. When hydrogen gas is heated to 10 or 100 million degrees, the particles that make it up move very rapidly indeed. Since hydrogen atoms cannot exist at such temperatures (any that form are quickly broken up), the gas consists of protons and electrons that are moving at velocities of thousands of kilometers per second.*

Now suppose that an astronaut steps out of his spaceship into this hot gas. Will he be vaporized by the high temperatures? No. Intergalactic gas is very thin; there are only a few particles per cubic meter. The few protons and electrons that strike the astronaut cannot possibly heat him to any appreciable degree. In fact, he will freeze solid if he is not wearing a spacesuit that has a heating unit. He will radiate heat away much faster than he will absorb it from the hot gas.

This illustration explains how the gas can maintain such a high temperature. Since it can heat objects only with great difficulty, there are no ways for it to cool down appreciably. The gas does convert some of its heat energy into x rays, but it does so very slowly. X rays are emitted when an electron passes by a proton and loses some of its energy in the encounter. But since the particles in the gas are so far apart, this rarely happens. As a result, the gas can remain hot for billions of years. It is perfectly conceivable that the hot gas has filled the spaces between galaxies since the galaxies were formed.

Astronomers have found some of the missing mass in clusters, but are not sure whether they have discovered all of it. The mass seen in the visible parts of galaxies is about

---

* Strictly speaking, only the components of hydrogen (protons and electrons) are present. However, physicists habitually refer to this mixture as *ionized hydrogen*. "Ionized" means that the electrons have been separated from the protons they had been circling.

10 percent of the amount that is needed to bind clusters together. The hot hydrogen gas contributes a roughly equal amount. It could be that the dark haloes make up the remaining 80 percent, but no one is sure. It has been determined that there is at least as much mass in the haloes as there is in the visible galaxies, but this is only a lower limit. It is perfectly possible that the missing mass might have some third, as yet undiscovered, component.

The concept of missing mass can also be applied to the universe as a whole. But here the term is used to designate something slightly different. The mass may not be invisible; it may be mass that really is missing.

Given the present rate of expansion, it can be calculated that the critical mass density for the universe is 5×10^{-27} kilograms per cubic meter. This translates to about a twentieth of an ounce per million billion cubic miles, or approximately three hydrogen atoms per cubic yard. If the mass density of the universe is less than this, then the universe is open; if the mass density is greater, then it is closed.

The "missing mass," in this context, is the difference between the mass that has been found and the amount that would be needed to close the universe. Here it is more accurate to say "missing mass" than "invisible mass," because no one is really sure whether or not it is really there.

If we take all of the mass that is concentrated in the visible parts of galaxies, we obtain a figure that is only about 2 percent of the critical amount. If we add in the mass that is known to exist in clusters, we have 10 or possibly 20 percent of the amount that is needed to close the universe. The remaining 80 or 90 percent is missing mass which may or may not exist.

The fact that there is dark, non-luminous matter in clusters of galaxies suggests, to some astronomers, that such dark matter may exist throughout the universe. If it does, then the average mass density could very well exceed the critical value. The universe, after all, is very big, and clusters occupy only a small part of the available space.

Astronomers have made numerous hypotheses concerning the forms the missing mass might take, and have made observations designed to see whether it is there. I will discuss the hypotheses one by one, considering the most plausible first.

The nearest large galaxy is M31, the great spiral galaxy in Andromeda. M31 lies at a distance of 2.2 million light years. By astronomical standards, it is very close.

Even at a distance of 2 million light years, it is not possible to make out many individual stars. Stars like the sun are not bright enough to be seen in the Andromeda galaxy. Even the most luminous supergiant stars cannot be detected at distances of more than 30 million light years. The only reason that we can see galaxies at all is that they are made up of billions of stars that are packed tightly together. Shining in unison, these stars manage to emit enough light to be visible.

Astronomers do not know how many stars exist outside of galaxies, or how many dwarf galaxies there are. The most powerful telescopes are not capable of detecting them unless they happen to lie very close to the Milky Way. If one is to determine whether or not such stars and small galaxies exist in significant numbers, it is necessary to use indirect methods.

Fortunately, there is a way of estimating the contribution made by such stars and galaxies to the average density of the universe. From this, we can determine whether or not it is possible that they might provide the missing mass needed to close the universe.

If there were a lot of individual stars scattered throughout the universe, they would produce light that could be seen as a faint background glow in the night sky, a kind of cosmic light. It so happens that there is a faint glow in the sky at night, but one cannot conclude that this is the cosmic light. First, one must subtract contributions from other sources.

Some of the background light is man made: Light from cities on the earth's surface is scattered by dust grains in the atmosphere. There is also *zodiacal light*, which is caused by interplanetary dust particles that scatter the sun's light toward the earth. While the zodiacal light is too dim to be seen in the daytime, it can easily be detected at night.

Finally, there is light from the Milky Way. Light from stars too far away or too dim to be seen contributes to the background glow, as does light scattered by interstellar dust.

The contributions from these sources account for all of the background light that has been detected. If the cosmic light exists it has not been seen, having been masked by the other sources.

But it is possible to calculate how much cosmic light any given number of stars or dwarf galaxies would produce. The fact that no light is seen allows us to place an upper limit on the number of such stars that could exist. If the stars are there, then they can provide, at most, 10 or 20 percent of the missing mass.

The next obvious place to seek the missing mass is in the

Missing Mass

space between clusters of galaxies. As we have already seen, there is gas between galaxies in clusters, so why could there not also be gas outside the clusters?

The idea is not an unreasonable one. The galaxies, after all, were formed from the primordial hydrogen and helium. If this process was not one hundred percent efficient, then quite a bit of this gas should be left over. Furthermore, there is no reason to think that it would all be found inside clusters.

In practice, one need only look for the hydrogen. If it is not present, there will be no helium. The helium, after all, was made out of hydrogen in the big bang, and there are no known processes operating in the universe that act to separate the two gases. Furthermore, there exist about 12 times as many hydrogen as helium atoms. The universe is roughly 25 percent helium by weight. But a helium atom is four times as heavy as one of hydrogen; 25 percent by weight is 92 percent in absolute numbers.

Hydrogen gas can exist in any of three different forms: It can be made up of hydrogen atoms; it can be composed of hydrogen molecules, in which two atoms are bound together; or it can be *ionized*, split apart into its constituent electrons and protons. The hot gas that exists inside clusters is ionized. So if we want to look for missing mass in the form of gas, we need only consider these three possibilities.

If the gas existed as hydrogen atoms, it would be easy to detect. Hydrogen absorbs certain wavelengths of electromagnetic radiation. One of these wavelengths lies in the radio spectrum, at 21 centimeters; another lies in the ultraviolet. The wavelength in the latter case is 1.216×10^{-5} centimeters, or 1216 angstroms. (An angstrom is a unit that is commonly used to measure wavelengths of visible and ultraviolet light; it is roughly equal to the diameter of an atom.)

If there is a significant amount of atomic hydrogen in the universe, it should absorb radiation from distant galaxies at these wavelengths. It turns out that there is some absorption, but not very much. Observations of 21-centimeter radio waves from quasars and radio galaxies allow an upper limit to be placed on the amount of atomic hydrogen that might be present. The total amount cannot be more than 20 percent of the critical density. Note that, as in the case of atomic light, the number is only a limit; there might be no atomic hydrogen at all.

A better estimate can be obtained by looking at the 1216-angstrom ultraviolet radiation. Since ultraviolet light of this wavelength is absorbed by the atmosphere, and cannot be seen from the ground, one must look at a distant object. Objects that are very distant have large redshifts. As a result, the ultraviolet radiation they emit is shifted into the visible parts of the spectrum.

An experiment performed by astronomers James Gunn and Bruce Peterson indicates that the amount of atomic hydrogen in the universe is less than one millionth of the critical value. This figure is almost too low; since the universe is made mostly of hydrogen, one would expect that more would be left from the era of galaxy formation.

But perhaps the hydrogen is there. It could be that it simply does not exist in the atomic form. Yet molecular hydrogen has not been found in significant amounts either. Studies of expected absorption effects (the method is the same, only the wavelengths are different) show that if hydrogen molecules do exist in intercluster spaces, the amount is less than one ten-thousandth the critical density.

That leaves only hot ionized gas. In the past, it appeared that hydrogen gas was present in this form, and that there might be enough to close the universe. The first astronomical x-ray observations, made from rockets in 1962, revealed the presence of an x-ray background. Like the microwave background detected by Penzias and Wilson, it came from every direction in the sky. It was believed that the existence of this background indicated that the universe was filled with ionized hydrogen at a temperature of 500 million degrees.

The controversy was resolved when detailed measurements provided by the x-ray telescope mounted in the HEAO-2 satellite indicated that the x rays were actually coming from distant quasars. Though this did not prove that no ionized hydrogen existed, it was apparent that there was not enough of it to be measurable.

The conclusion that must be drawn is that the missing mass cannot exist in the form of hydrogen gas. The hydrogen in the spaces between clusters of galaxies can provide, at most, a few percent of the critical mass density. If the missing mass is present, it must take some other form.

If we cannot find the missing mass in the form of stars, dwarf galaxies, or gas, what about solid, non-luminous matter? Such matter could conceivably exist in many forms, and range in size from dust particles to planets. The intercluster spaces could even contain dead, burned-out galaxies.

If dust particles existed in large enough numbers, they would scatter the light that comes to us from distant galaxies. The universe would then seem to be polluted with a kind of cosmic smog; distant objects would have an appear-

ance as hazy as that of distant buildings in a city with polluted air.

The other possibilities cannot be disproved, but are
thought to be unlikely. Though pebbles, rocks, planets,
and even "snowballs" of frozen hydrogen could exist unde-
tected throughout the universe, astronomers have difficulty suggesting their possible origin. The existence of
dead galaxies is also improbable. Stars the size of the sun
have lifespans that are not much shorter than the present
age of the universe, while stars smaller than the sun can
live considerably longer. The idea that there should exist
galaxies that have grown completely dark does not sound
very reasonable.

But what about radiation? According to Einstein's fa-
mous formula, $E = mc^2$, energy has mass. If enough mass
existed in the form of radio waves, visible light, infrared
and ultraviolet radiation, and gamma rays and x rays, it
could conceivably close the universe.

In the early stages of the big bang, radiation constituted
the dominant part of the mass of the universe. However,
the radiation-dominated era of the universe came to an end
billions of years ago. Today the mass density of radiation is
almost a hundred thousand times smaller than the mass
density of matter.

Cosmic rays cannot provide the critical density either;
they can provide, at best, only about a thousandth of the
necessary mass.

If we are to believe in a closed universe, three reasonable
possibilities remain. The first is that the missing mass
exists in the form of black holes. The second is that it consists of a sea of neutrinos. The third is that the missing

mass is of the same form as the galactic haloes, whatever that might be.

Neutrinos and black holes will be discussed in more detail in later chapters. For the moment, I will note only that there is no way of telling how many of either the universe might contain. Since black holes emit no light, they can only be observed when their intense gravitational fields affect nearby objects. It is equally conceivable that black holes might not exist or that they might exist in such great numbers as to provide the critical mass density many times over. And the situation with respect to neutrinos is even more uncertain.

There remains the question of galactic haloes. Since we do not know how far these extend, it is not possible to say what fraction of the missing mass they provide. Measurements of masses in clusters are insufficient because evidence indicates that entire clusters could have haloes. If this is true, cluster haloes could contain enormous quantities of matter.

Alternatively, the universe could contain dark isolated condensations of matter. Since it is not known what haloes are made of, it is impossible to say that similar dark clumps of matter cannot exist; the universe could be full of massive halo-like "galaxies" which emit no light.

These three possibilities are not mutually exclusive. The halo could partly consist of neutrinos or black holes. Another possibility is that it is composed of very dim stars.

Yet another possibility is that haloes are made up of bodies too small to ever have become stars. Although theories of star formation have not been worked out in as much detail as astronomers would like, it is known that a certain minimum mass is required. If a contracting gas cloud does not have the minimal amount of mass, then gravitational attraction can never heat it up enough to start the nuclear reactions that take place in a star's core. The planet Jupiter provides an excellent example of such a "failed" star. If Jupiter had only a little more mass, it would have begun to shine at about the same time that the sun did, and there would be two suns in earth's sky.

Note that the hypothesis that such bodies make up the material in haloes is not the same as the theory that they could be scattered throughout intergalactic space. Although the latter theory was discounted previously, the arguments against it do not necessarily apply in the case of haloes.

But all this is speculation. Until astronomers find out what haloes are, they will have no way of telling how much matter exists in this form.

It should be obvious by now that the fact that the missing mass has not yet been found does not imply that it is not there. The arguments concerning the abundance of helium and deuterium seem to imply that it should not be, yet those arguments could easily be overthrown. They are only theoretical, after all. If observations indicated that there were indeed slightly more than three hydrogen atoms for each cubic meter of the universe, it would be established that the universe was closed.

But of course that "little more" may never be found. The universe could turn out to be open after all.

CHAPTER 5

Black Holes

WHEN Gott, Gunn, Schramm, and Tinsley published their paper on the open universe in 1974, they expressed their conclusions in a cautious scientific manner. "A variety of arguments," they said, "strongly suggest that the density of the universe is no more than a tenth of the value required for closure." But, they hastened to add, the possibility that the universe might be closed was not entirely ruled out. The arguments against a closed universe, they judged, were "formidable, but not fatal." The four authors even suggested the form that the missing mass might take. At the conclusion of a statement which indicated their belief that the universe was probably open, they added punningly, "Loopholes in this reasoning may exist, but if so, they are primordial and invisible, or perhaps just black."

Although the topic of black holes is one of the most widely discussed subjects in contemporary astronomy, the idea that a body might have a gravitational field so strong that it could not emit any light is not a new one. It was first suggested in 1784, when the English astronomer John Michell published an article in *Philosophical Transactions of the Royal Society of London* which suggested that if a star was massive enough, it could trap all the light it emitted. The gravitational attraction would be so strong that nothing, not even light, could escape from its surface.

Twelve years later, the French mathematician Pierre Simon de Laplace, who was apparently not aware of Michell's work, put forth similar arguments. The only significant difference between Michell's theory and Laplace's was that Michell calculated that a star would have to be 497 times larger than the sun if it was to be a black hole, while Laplace came up with a figure of 250.

Although the idea turned out to be remarkably prophetic, the details of the two scientists' calculations were, of course, incorrect. Both men used Newton's law of gravitation, the only gravitational theory that was known to them. Today we know that Newton's theory cannot accurately describe the intense gravitational fields that exist in and around a black hole; Einstein's must be used if one is to have any hope of obtaining a correct result.

Contrary to what Michell and Laplace believed, a black hole is not a big star. No star, however massive, can produce gravitational forces of the required intensity. If a black hole is to form, the matter that makes up the star must somehow be squeezed into a very small volume. This can only happen when a very massive star collapses after exhausting its supply of nuclear fuel. Black holes, in other words, are the dead relics of stars that have been compressed by their own gravity.

Before the death of stars can be discussed, it will be necessary to make a few remarks about how they are formed. Stars are still being born within the Milky Way, and presumably in most of the other galaxies. We can observe star formation in our own galaxy, and there is no good reason for thinking that conditions elsewhere should be any different.

Clouds of gas are found scattered throughout the Milky Way. According to currently accepted theories, shock waves associated with the moving spiral arms, or with supernova explosions, can cause these clouds to undergo a slight contraction. This is all that is needed to trigger the process of star formation, for once the contraction begins, nothing can stop it.

As the cloud contracts, it fragments into thousands of smaller clouds. Gravity causes these to contract still further. Compression causes the gas within them to heat up, and this slows the process but does not stop it, for most of the excess heat is simply radiated off into space. After millions of years, gravity causes the fragments to condense into hot, glowing *protostars*.

If the mass of a protostar is more than a few percent of the mass of the sun, then the temperature and pressure in its core will eventually become high enough to begin the nuclear reactions that will transform it into a star. If the mass is less than this critical amount, these reactions will never get started, and the protostar will cool into a *black dwarf*.

During the earlier stages of their lives, all stars transform hydrogen into helium. This process of *thermonuclear fu-*

sion is fundamentally the same as that which takes place within a hydrogen bomb explosion. The only important difference is that an H bomb burns up all its fuel in a fraction of a second, while a star can go on shining for billions of years.

When hydrogen is converted into helium, mass is transformed into energy according to Einstein's equation $E = mc^2$. The mass lost is small. It takes 4.0325 grams of hydrogen to make 4.0039 grams of helium, so the deficit amounts to only 0.0286 grams.

But large amounts of energy can be created from small amounts of matter. For example, the sun has enough fuel to continue shining for another 5 billion years, at which time it will be about 10 billion years old. Yet during its entire life it will convert only 0.01 percent of its mass into energy. A glance at Einstein's equation will explain how this can be. The velocity of light, c, is a very large number. In metric units, it is about 300,000,000 meters per second. Therefore c^2 is 90,000,000,000,000,000, or 9×10^{16}. This tells us that 1 unit of mass will produce 9×10^{16} units of energy.

The conversion of hydrogen into helium is the process that produces the heat and light that a star radiates into space. But thermonuclear fusion also performs another very important function; it keeps a star from collapsing.

The temperature in the core of a star, where the fusion takes place, is about 15 million degrees in the case of a star the size of the sun. The core is, as we have already seen, made up of hydrogen and helium gas. Now when a gas is heated, pressure is created. This is what happens in an automobile engine, for example. When the gasoline-air mixture is ignited, the combustion creates heat, and the resulting pressure drives the piston. It should be obvious that if a temperature of a few hundred degrees can power a car, then a temperature of millions of degrees should be sufficient to keep the outer layers of a star from falling into the core. Naturally, gravity prevents the material from being blown off into space.

One would expect that when a star's hydrogen fuel runs out, it should begin to shrink. This is exactly what happens. But the shrinking does not begin at once; things are not quite that simple. Before a star can get smaller, it must first become larger. As stars begin to die, they expand into *red giants*.

Astronomers do not yet understand all the details of stellar evolution during the red giant stage. But it is clear that this phase begins when the hydrogen in a star's interior is exhausted. Once this happens, the nuclear reactions stop. Gravity causes the core to contract. The contraction, in turn, causes the core to heat up. But as this happens, the outer layers of the star expand and become cooler. The lowered temperatures bring about a change in the star's color and it turns red. The change is analogous to those which are observed in a cooling white-hot iron bar; as its temperature drops, its color changes from white to yellow to red.

The sun will become a red giant in about 5 billion years. As it goes through this transformation, it will expand until its surface lies at about the present orbit of Mars. The inner planets—Mercury, Venus, Mars, and the earth—will be vaporized. During the red giant stage, the sun will eject some mass; dying stars typically go through various kinds of spasms. But eventually the core of the sun will cool and the outer layers will shrink. The sun will gradually contract and grow dimmer until it becomes a *white dwarf*. At this point, contraction will halt.

White dwarfs are quite common in our galaxy. Tens of thousands of stars suspected of being white dwarfs have been catalogued; astronomers have studied more than five hundred of these quite closely. White dwarfs are tiny, extremely dense stars. They are only about the size of the earth, and they are so condensed that a matchbox full of white dwarf matter would weigh about ten tons. No nuclear reactions take place within white dwarf cores. These stars shine only because they are so hot; it takes billions of years for a white dwarf to cool and to fade into darkness.

But it is not the pressure created by the residual heat that holds up the surface of a white dwarf. This is done by the electrons that are one of the constituents of white dwarf matter. When electrons are packed very tightly together, they begin to exert a pressure which resists further compression. The technical name for this is *electron degeneracy pressure*. It is called this because electrons packed together in this manner are said to be in a "degenerate state."

The material that makes up a white dwarf bears little resemblance to ordinary matter. Atoms and molecules do not exist; instead, there is a sea of closely packed electrons in which the atomic nuclei float. Matter in such a state is called a *degenerate gas*.

Most stars end their lives as white dwarfs. But if a contracting star has more than 1.4 solar masses, it will experience a different fate. The gravitational forces in a star this large are strong enough to overcome the electron degeneracy pressure. The electrons and the protons within the nuclei are squeezed together, forming neutrons. As soon as this happens, the degeneracy pressure disappears, for the simple reason that the electrons no longer exist.

Every star contains equal numbers of electrons and protons. All matter is balanced in this way. A hydrogen atom is made up of an electron and a proton. In helium, two electrons circle a nucleus that contains two protons and two neutrons. An atom of uranium 238 contains 92 electrons, 92 protons, and 146 neutrons (the number 238 is the sum of 92 and 146, the total number of particles in the nucleus). If all of the electrons and protons are combined, nothing but neutrons will be left. A star that contains only neutrons will contract until the neutrons are packed together.

A *neutron star* is said to be held up by *neutron degeneracy pressure*. Neutron stars are much smaller than white dwarfs; they are typically about 20 kilometers in diameter (roughly the same as the length of the island of Manhattan). They are about a billion times denser than the white dwarfs; a matchbox full of neutron star matter would weigh about 10 billion tons.

It is believed that most or all neutron stars are formed in the aftermath of supernova explosions. A red giant massive enough to evolve into a neutron star does not do so quietly. When the nuclear fuel is exhausted, it undergoes a cataclysmic explosion that blows off the star's outer layers. Meanwhile, the core collapses rapidly, passing quickly through the white dwarf stage and becoming a neutron star as electrons and protons are fused together.

Supernovae, incidentally, are not related to ordinary novae. The latter are, by astronomical standards, relatively mild events that can take place thousands of times in the same star. It is believed that novae take place only in binary star systems. They seem to be flareups that happen when the gravity of one star draws material from the surface of its companion. As this material falls onto the more massive of the two stars, a small amount of energy is released, and the star system increases in brightness. After a time, the flareup dies away, only to be repeated at a later date. A nova has the appearance of a bright star, while a supernova can shine with the brightness of an entire galaxy.

Some neutron stars contain magnetic fields which cause them to emit light or radio waves in certain directions. The beams created rotate through the sky in the manner of a searchlight, and this radiation has the appearance of a series of pulses. Such neutron stars are therefore called *pulsars.*

All stars spin, but neutron stars rotate especially fast; their spin increases dramatically when they collapse into the neutron star state. The effect is analogous to that which takes place when a spinning ice skater brings her outstretched arms down to her sides to make herself rotate faster.

The first observation of a pulsar was made by the English radio astronomer Jocelyn Bell in 1967. The pulsar that Bell discovered produced radio pulses 0.016 seconds in length once every 1.337 seconds. The following year it was established that this pulsar lay in the middle of the Crab Nebula, the remnant of a supernova explosion that was observed on the earth in A.D. 1054 by Chinese astronomers. Since 1968 a number of other pulsars have been discovered. Some of these emit light, while others are observed by their radio emissions. The case for the existence of neutron stars can therefore be considered conclusive.

If gravitational fields become strong enough, they

should be able to cause contraction beyond the neutron-star stage. Scientists are not sure exactly how much mass a star must have before this happens, but it is probably safe to say that the upper limit to neutron star size lies somewhere between 1.7 and 2.7 solar masses.

If this limit is exceeded, the gravitational force crushes the neutrons together. The theory of general relativity tells us that nothing can stop the collapse; it continues until all matter in the star is compressed into a mathematical point of infinite density, a black hole.

Although all of the matter in a black hole is compressed into the singularity at the hole's center, it can still be said to have a finite size. Every black hole is characterized by a surface called the *event horizon.* Anything that enters the hole by crossing the event horizon is forever trapped; it can never recross the surface in the other direction. It will be inexorably drawn toward the singularity, the region of infinite density in the center of the black hole.

In a way, a black hole is a small replica of a closed universe. The gravitational fields within the hole curve space so strongly that space closes upon itself, cutting the interior of the black hole off from the rest of the universe. The fact that nothing can get out of a black hole is analogous to the fact that it is not possible to get "outside" of the universe, even though it might be finite.

It is easy to see where the term black hole comes from: These objects are blacker than anything else encountered in nature, for they neither emit nor reflect any light. If a chunk of matter that had passed through the event horizon did give off light rays, they would be pulled back by the enormously strong gravitational fields, and would follow the matter as it fell into the singularity. If a spaceship could somehow shine a searchlight on a black hole, the occupants of the ship would see nothing; the black hole would absorb all the light that fell on it.

Are there really singularities in the centers of black holes? General relativity says there must be. But can we really expect that the theory will remain valid under such extreme conditions? Isn't it possible that future generations of physicists will discover new physical laws which will show that the mass density in the center of a black hole, although very large, is not actually infinite?

These questions were discussed in a different context in Chapter 3, where it was noted that general relativity also predicts infinite density at the beginning of the big bang. It was also noted that as long as we lack a theory of quantum gravity, it is not really possible to extrapolate all the way back to the beginning.

A similar situation exists with respect to black holes. We cannot be sure that black hole singularities really exist. It could very well turn out that some process prevents matter from being crushed into a mathematical point. Naturally, no one knows what such a process would be.

General relativity may not be valid in the center of a black hole. However, it should describe the initial stages of black hole collapse perfectly. It is possible to say with some confidence, therefore, that if there are any collapsing stars with more than 2.7 solar masses, then black holes must inevitably form. We may not understand all the details of the process that takes place afterward, but we do know that if there are dying stars of this size, then black holes must exist.

A number of stars have been observed in the 40 to 50 solar mass range, and there may be others that are even larger. In 1981 three University of Wisconsin astronomers announced that they had discovered a star that was more than 3,000 times as massive as the sun.

But it does not necessarily follow that stars are very large at all by the time they undergo their final collapse. Astronomers are not sure how much mass a star can eject during the red giant stage, or how much is blown away in a supernova explosion. They are not even sure whether black holes do form in the aftermath of supernovae the way that neutron stars do. They know of no mechanism which would allow a star of 40 or 50 (or 3,000) solar masses to get rid of enough mass to avoid the fate of black hole collapse. But this does not mean that it could not be done.

In other words, it is not enough that theoretical arguments say that black holes should exist. If we want to find out whether they do exist we must find a way to observe them. If one black hole can be found, we will know that black hole formation is possible, and can reasonably assume that many are scattered throughout the universe.

But how is one to observe something that cannot be seen? This is not so difficult as it sounds. Physicists know of many objects that cannot be observed directly. For example, no one has ever photographed any of the various kinds of subnuclear particles; they are far too small to be seen. But it is possible to observe the nuclear reactions they enter into.

The situation regarding black holes is similar. If they exist, black holes will interact with visible bodies. If these interactions can be observed, then it can be established that black holes are real objects.

In particular, if a black hole is very close to a normal star, then the interaction between the two should produce x rays. The intense gravitational field of the black hole will pull gas from the surface of the star.* As this gas spirals into the black hole, it will become very hot and dense. The nearer it gets to the event horizon, the hotter it gets. When it becomes hot enough, it will begin to emit x rays. If the black hole and its companion are close enough to us, it should be possible to observe these x rays from the earth.

Since about 50 percent of all star systems are binary, it follows that black holes must frequently have companions that are visible stars. But it does not follow that all of these systems will produce x rays. In fact, most of them will not. In some cases the visible star and the black hole will be too far apart. In others, the x-ray emission will not have started, or will have stopped. There might even be other, as yet unknown reasons why x-ray emission might not be observed in any particular case. However, if there really are such things as black holes, then somewhere among the billions of star systems that make up the Milky Way galaxy, we should be able to see the kind of x-ray emission that is predicted.

Although the existence of black holes was predicted on theoretical grounds as long ago as 1939 by the American physicist J. Robert Oppenheimer and his student Hartland Snyder, thirty years were to pass before it became possible to look for them. The problem was that x rays coming to us from outer space cannot be detected from the surface of the earth; it is necessary to make observations from a spot above the atmosphere.

This became possible in 1969, when the satellite Uhuru was put into orbit. *Uhuru* is the Swahili word for "free-

* The orbital motion of a star keeps it far enough away from the black hole that there is no danger that it will be gobbled up too.

98

dom"; the satellite was so named because it was launched from the coast of Kenya on the fifth anniversary of Kenyan independence. Though x-ray observations had previously been made with equipment placed on rockets, these experiments were unsuccessful. Rockets fall back into the atmosphere after a short time, making detailed study of x-ray sources impossible.

Uhuru discovered about 200 x-ray sources. But this does not mean that it discovered 200 black holes; x rays can be produced in numerous ways. They are emitted by gas clouds, by neutron stars, by quasars, and by hot white dwarfs. X rays are even produced by ordinary stars. However, theoretical calculations indicate that x rays from black holes should flicker. This flickering is caused by the fact that matter is not drawn from the hole's companion in a steady stream—there are irregularities in the flow.

In 1971, Uhuru discovered such a flickering source in the constellation Cygnus ("the Swan"). Named Cygnus X-1 (x-ray source number 1), it seemed to be a likely black hole candidate. However, scientists were not willing to jump to any conclusions. At this point, it had not even been shown that Cygnus X-1 was located in a binary system. And once that was done, it would still be necessary to establish that the bodies in the system had masses of the right order of magnitude. For example, if it turned out that they were both too small to be black holes, then the whole hypothesis would go down the drain.

During the next few years, an enormous number of astronomical observations were made. First, Cygnus X-1 was identified with a visible star, a blue giant known as HDE 226868 (number 226,868 in the extension of the Henry Draper catalogue). It was established that this young, hot, blue star was about 25 times as massive as the sun, and that it had an invisible companion between 8 and 18 solar masses. The two bodies revolved around one another once every 5.6 days. The shortness of this time indicated that they had to be very close together.

According to currently accepted theory, any dark body that has 8 or 18 solar masses must be a black hole. The upper limit on neutron star mass is, as we have seen, somewhere between 1.7 and 2.7 solar masses. The figure of 8 is far enough above that so that even if there are some inaccuracies in the theory, the body should still be a black hole.

Throughout the 1970s various groups of astronomers tried to invent explanations for the x rays from Cygnus X-1 that did not depend on the presence of a black hole. Cygnus X-1 might contain three bodies, none of which is a black hole, they said. Or possibly Cygnus X-1 contained two magnetized stars; it could be that the magnetic fields were responsible for the observed radiation. Or maybe HDE 226868 wasn't as far away as it seemed to be. If it was much closer than astronomers believed, then it would be necessary to lower the estimates of the masses of both HDE 226868 and its companion. And if this was done, the latter would almost certainly lose its presumed black hole status.

Every one of these proposals was considered carefully. Theoretical scrutiny and additional astronomical observations showed, however, that all were wrong, or at least very implausible. At the present time, there seems to be no way to deny the fact that Cygnus X-1 probably contains a black hole. However, I think that most astronomers would want to retain the word "probably" in such a statement. Until the existence of black holes is confirmed in some other way, it is not possible to be absolutely certain.

There are now a number of other black hole candidates besides Cygnus X-1. One of these, known as V 861 Scorpii, seems especially promising. But it is not enough to collect a lot of examples of promising possibilities. If the existence of black holes is to be demonstrated beyond any reasonable doubt, it will be necessary to observe them by another method.

Such confirmation of the theory could be obtained in any of a number of different ways. If the attempts to detect gravitational waves are successful, then gravitational-wave detectors might be used to search for black holes, because gravitational radiation should be produced when a black hole has a close companion.

The gravitational-lens effect suggests another possibility. A massive black hole could bend the light from distant objects the same way a galaxy does. But the fact that a galaxy is thousands of light years across, while a black hole has a very small volume, implies that the images should be somewhat different in the two cases. It might be possible to analyze a gravitational lens and to use the data to deduce the presence of a black hole.

The evidence provided by gravitational-wave or gravitational-lens observations might not be quite as convincing as that obtained from x-ray observations of Cygnus X-1. Nevertheless, the case for the existence of black holes would be made stronger.

If black holes exist, could they provide the missing mass needed to close the universe? No one really knows the answer to this question. However, theoretical arguments indicate that black holes are not numerous enough to provide the missing mass.

The first such argument is based on currently accepted ideas about the formation and evolution of stars. If current theories are reasonably correct—and there is every reason to think that they are, in spite of the fact that some difficulties remain in the theory of star formation—then it is possible to calculate how many stars should have collapsed into black holes since the beginning of the universe. It turns out that about 1 percent of all stars should have undergone such a collapse. Since the stars do not contain enough mass to close the universe, the 1 percent that may have become black holes most certainly could not do so either.

The second argument has to do with the observed helium and deuterium abundances of the universe. As I pointed out in Chapter 3, the measured amounts of these two substances indicate that the density of matter in the universe is much less than that required for closure. It should not make any difference whether this matter exists in the form of gas, as galaxies and stars, or as black holes. Black holes, after all, seem to be made of the same material as everything else.

There are loopholes in these arguments. The helium-deuterium argument has implications only for the density of ordinary matter, including matter that has condensed into black holes. But if the universe were largely made up of something that is not ordinary matter, it could very well be closed after all. Although such an idea would have been dismissed as the purest fantasy a few years ago, new discoveries have suddenly made it very hard to discount. These will be discussed in detail in the next chapter.

The other possible loophole has to do with the possible existence of *primordial black holes* (often called mini black holes, or miniholes). The helium-deuterium argument might not apply in this case, for the primordial black holes could very well have been made first. Furthermore, theory does not give us an estimate as to how many there might be, because they are not collapsed stars.

In 1971 the English physicist Stephen Hawking pointed out that very small black holes could have been created early during the history of the universe. Shortly after the big bang, the universe was very dense. Fluctuations could have acted to condense the material in the fireball to even higher densities in spots, thereby forming a large number of black holes. Although it is thought that they would have been formed in varying sizes, most would be smaller than an atomic nucleus (and yet would weigh more than a mountain).

One might think that it would be impossible to detect such objects. After all, it is hard enough to observe the very massive black holes that are formed from collapsed stars. But this does not turn out to be the case. Hawking has pointed out that very small black holes should spontaneously explode after the passage of a certain period of time. For a black hole with a mass of 10^{15} grams (about a billion tons), this time turns out to be roughly equal to the age of the universe.

Therefore, if primordial black holes of this size do exist, we should be able to see them blowing up right now. When they explode, large quantities of gamma rays should be emitted, and it should be possible to detect these from the earth.

Although such gamma rays have not been observed, it

does not necessarily follow that primordial black holes do not exist. But if they do, there cannot be very many of them. At most there are only a few thousand per cubic parsec (a parsec is 3.26 light years; a cubic parsec is about 35 cubic light years). The resulting mass density turns out to be only about a hundred-millionth of that required to close the universe.

One must conclude that it is not very likely that primordial black holes could play much of a role in halting the present expansion. It has not been established that they really exist. And if they do exist, there cannot be very many of them. Unless, of course, they were formed in such a way that they all weigh more than 10^{15} grams. Theory, however, says that this is very unlikely. Their masses should vary in a random manner, related to the random fluctuations that were presumably responsible for their formation.

Confirmation of the probable existence of black holes has led to a lot of speculation, some of which not even scientists take very seriously. However, there are some very fantastic-sounding suggestions that are not so easy to dismiss.

The idea that has received the greatest amount of publicity is the suggestion that black holes might function as gateways to other universes, or to our own universe at some other point in space and time, possibly billions of years in the past.

It must be emphasized that contrary to what some popular books imply, most astronomers view this theory with extreme skepticism. However, even the confirmed skeptics are willing to admit that this is a legitimate area for scientific speculation. After all, we would never have progressed as far as we have if scientists had a habit of refusing to consider ideas that sounded fantastic. Anyone who is at all familiar with the history of modern physics should realize that it is precisely the "fantastic" ideas which often have the best chance of turning out to be true.

In an earlier part of this chapter, I left the reader with the impression that anything that enters a black hole must be crushed out of existence in the singularity that lies at the hole's center. If the black hole is static, this is perfectly true. But if the black hole is rotating, it is not necessarily the case.

Theoretical calculations indicate that the singularity at the center of a rotating black hole does not take the form of a point, but of an infinitely thin ring. The theory says, furthermore, that it should be possible for an object to fall into the black hole in such a way that it will avoid a collision with the ring. If it is pointed in just the right direction, it can miss the singularity by passing through the middle of the ring.

An object that does this enters some previously unknown region of space. However, the theory tells us nothing about what this region of space is like. It could be another universe with unknown properties and physical laws. Or it could be our own universe at a time shortly after the big bang.

The theory of black hole gateways is based on general relativity, specifically on the prediction that there will be a region of infinite density in the center of every black hole. But general relativity breaks down in regions of high density. Until a theory of quantum gravity is developed, we cannot be sure that theory is giving us an adequate description of what really does happen in the center of a black hole.

If we cannot be sure that there are singularities, then we obviously cannot have a lot of confidence in ideas about other universes. Furthermore, certain theoretical arguments indicate that even if the gateways are real, they close off as quickly as they open, with the result that it is impossible to get through them.

If other universes do exist, it is useless to ask where they might be. The question probably does not even have any meaning. We can only ask "where" something is if it exists in our own space and time, and hypothetical alternate universes obviously do not. It is equally useless to ask how many of them there are. If they really do exist, their number might be infinite.

Such ideas are so intriguing that we sometimes find ourselves believing in them whether there is any good reason for doing so or not. So perhaps I should emphasize again that the concept of black holes as gateways to other universes is based upon an extrapolation of the theory of general relativity into a region where the theory is known not to be valid.

But, for better or for worse, black hole theory has had the effect of making speculation about other universes scientifically respectable. The topic of "parallel universes" is no longer purely science fiction. During the last few years, it has become part of contemporary cosmological thought.

The theory of black holes has led to another fascinating suggestion. Let us assume, for a moment, that our universe is closed. If it is, it resembles a black hole in certain important respects: Nothing that enters the event horizon of a black hole can escape; nothing can get outside of a closed universe. There is a singularity at the center of a black hole; a closed universe must collapse to a singularity at some time in the future. A black hole separates itself from the rest of the universe by curving space so strongly that space closes in upon itself; a closed universe curves space in the same manner.

Such considerations have led to the speculation that our universe is a black hole within a larger universe. It is true that a closed universe has no boundary, while a black hole has an event horizon, but this is not a good argument against the idea. Since nothing inside a black hole can get back out to the event horizon, the horizon has, for all practical purposes, ceased to exist. If a closed universe had an event horizon too, it could only be seen from the larger universe that contained it.

Admittedly, the idea is a little hard to take. If one is going to assume that our universe is a black hole within a larger universe, then there is no good reason for rejecting the idea that the latter may be a black hole within a still larger universe. Such a progression could go on forever— one could have an infinite number of universes nested within one another like Chinese boxes. There is nothing about this that is impossible. Yet the idea of an infinite series of universes that grow infinitely large does have an air of implausibility about it.

Is it really any worse than the concept of an open universe, which is also infinite? No one knows. And that is the trouble with speculation of this sort. Only experiments and scientific observation can tell us what is real and what is not real. But the minute we begin speaking of other universes, we seem already to have gone farther than observation and experiment can follow. When that happens, it is not easy to tell to what extent one has severed contact with reality. In fact, one begins to wonder whether "reality" is such a concrete thing as one has always assumed it to be.

CHAPTER 6

Elusive Particles

By the end of the 1970s, most astronomers had become convinced that the universe was open. The evidence seemed to allow no other conclusion: The helium-deuterium arguments were just too convincing, and the missing mass had not been found.

Then in 1980 new discoveries were made that indicated it was possible that most of the mass in the universe might be something other than ordinary matter. If this was the case, the universe could be closed after all. Suddenly there appeared to be loopholes in the most convincing arguments for an open universe.

Strictly speaking, the arguments based on the observed quantities of helium and deuterium do not place limits on the amount of mass that can be present in the universe; they apply only to matter that is made out of *baryons*.

"Baryon" is simply the physicist's term for a heavy particle. The most common baryons are neutrons and protons.
Although many other varieties exist, I will not say anything
further about them. They are generally observed only in
experiments performed with high-energy particle accelera-
tors, and are not very important to cosmology.

When the primordial helium and deuterium were made,
it was still much too hot for electrons to attach themselves
to nuclei. And so if we want to be precise, we should say
that it is helium and deuterium *nuclei* that were made in
the big bang fireball. The electrons that were later to combine with them to form helium and deuterium *atoms* were
still moving in every direction at high velocities.

The formation of helium and deuterium nuclei was a
process that involved baryons—neutrons and protons—
only. Therefore all that the theoretical arguments concerning their abundance really tell us is that there are not
enough baryons to close the universe. The theory says
nothing about other forms of matter.

Until recently it was believed that baryons constituted
almost all of the mass in the universe, so no one bothered
much about the distinction. The electron weighed only
1/1836 as much as a neutron or proton. For all practical
purposes its mass could be neglected. The other particles,
such as *mesons*—the nuclear "glue" that binds neutrons
and protons together—and *muons*, or heavy electrons, did
not exist in great enough numbers to be important. To say
that there were not enough baryons to close the universe
and to say that there was not enough matter seemed to be
the same thing.

Physicists were aware that the universe contained something like 10 billion neutrinos for every atom of ordinary

matter, that there were about 450 of them for every cubic centimeter of space. However, the neutrino seemed to have no mass, which made it a poor candidate for closing the universe. It did have energy, of course, and energy is only mass in another form. But calculations showed that a massless neutrino should contribute even less to the average mass density of the universe than radiation does.

In 1980, it was discovered that this elusive particle might have mass after all. If it did, then it was obvious that the greater part of the mass in the universe might very well be concentrated in neutrinos.

In other words, neutrinos could very well be the major constituent of the universe. The stars and the planets, the galaxies, quasars, and black holes, the interstellar gas and dust might be relatively unimportant; they might be nothing more than impurities floating in a sea of ghostly particles.

Neutrinos rarely interact with ordinary matter. It is estimated that billions pass through our bodies every second, yet only a few will interact with any of our constituent atoms during a lifetime. The penetrating power of these particles is so great that a typical neutrino could easily pass through a block of lead stretching from the earth to the nearest star.

Neutrinos are odd particles, and they have a strange history. In 1897, shortly after the discovery of radioactivity, the English physicist Ernest Rutherford began studying samples of the element uranium. A long series of experiments performed the following year showed that uranium emitted two distinct kinds of radiation, which Rutherford called *alpha* and *beta* after the first two letters of the Greek alphabet. Shortly afterward, the French physicist P. V.

Villard found a third, even more penetrating kind of radiation, which was naturally called *gamma*.

Gamma rays are a form of high-energy radiation. But alpha and beta "rays" are not. On the contrary, they are tiny chunks of matter. An alpha particle is made up of two protons and two neutrons, as is a helium nucleus. A beta particle is simply an electron.

Of course Rutherford didn't know this. In 1898, no one knew what electrons and protons were. Scientists had no idea what atoms looked like; many even doubted that they really existed. Thirteen years were to pass before it was shown—by Rutherford, again—that an atom was made up of a small, positively charged nucleus that was orbited by negatively charged electrons.

By the beginning of the 1920s, these matters were well understood. Or so scientists thought. A series of experiments had unraveled the secrets of atomic structure, and the nature of alpha and beta particles had been determined. The one major exception was the process by which beta particles were emitted. The more physicists studied beta decay, the more puzzled they became. Finally they had to admit that they had encountered something that they did not understand.

In order to see their problem, we will look at the simpler alpha decay process first. Numerous different kinds of radioactive atoms can throw off alpha particles. One of these is the atom that Rutherford studied, uranium 238. When a uranium atom emits an alpha particle, it loses two of the protons and two of the neutrons that are contained in its nucleus, and is transformed into thorium 234.

If we add up the masses of the alpha particle and the thorium 234, we obtain a result that is slightly less than the mass of uranium 238. There is nothing very surprising about this; it is simply an indication that a small amount of mass has been transformed into energy. If this did not happen, there would be no such thing as alpha decay. The alpha particle, after all, must have some energy of motion, or it could not be emitted in the first place.

The alpha particle's energy of motion is equal to the mass deficit. One need only apply the equation $E = mc^2$ to see that all the mass that disappears turns up again in the form of energy.

But things do not seem to balance out this way in the case of beta emission. When a beta particle is emitted by a radioactive atom—for example, thorium 234—it does not always come off with the same energy. The energy of the particles that are emitted varies continuously over a fairly wide range.

This is very surprising. The amount of mass that is lost is always the same, just as it is in the case of alpha decay. But as far as the physicists of the 1920s could determine, not all of this mass turned up again as energy. Some of it disappeared. And, to make things worse, the amount that disappeared was not always the same. Sometimes the beta particle seemed to have all of the energy of motion that it should have. But, on the average, about half of it was missing.

Conservation of energy is one of the fundamental laws of physics. It says that energy simply does not disappear. It can be converted into mass, or it can be changed into another kind of energy, but it does not vanish. For example, the electrical energy that enters a television set is transformed into light, sound, and heat. The chemical energy in gasoline can be changed into the energy of motion that is needed to run an automobile engine. Plants can make use of the process of photosynthesis to store radiant energy from the sun. We are able to store the energy we get from plant and animal foods in our bodies.

Physicists had never found any exception to this rule. But in the case of beta decay, either energy was not being conserved, or a small amount of mass was simply disappearing.

In 1930 the Austrian physicist Wolfgang Pauli suggested a possible solution to this problem. Suppose, Pauli said, a second particle was emitted at the same time as the beta particle. If the available energy could be divided between the two particles in various different ways, this would explain why the beta particles were less energetic than they apparently should be.

Since the new hypothetical particle was electrically neutral (it had no electric charge), Pauli called it the "neutron." But it did not keep this name for long. Two years later, in 1932, the English physicist James Chadwick discovered the particle that is now known as the neutron. Since it too was electrically neutral, Chadwick (who was apparently not familiar with Pauli's work) chose the same name.

This problem in nomenclature caused some confusion until the Italian physicist Enrico Fermi cleared everything up by explaining to seminar participants in Rome that Pauli's particle was only a *neutrino* (Italian for "little neutron"). The name stuck, and the particle has been called the neutrino ever since.

Some physicists were not too happy with the idea of the neutrino, for it seemed to be a rather ghostly particle. If Pauli's hypothesis was correct, the neutrino would have to be a massless particle that traveled at the speed of light. It could apparently not be detected, yet it did have energy, and spin as well. If some physicists considered it to be roughly analogous to the grin of the Cheshire Cat, they cannot be blamed for their skepticism. Even Pauli seemed to experience a few pangs of conscience. "I have done the worst thing for a theoretical physicist," he confided to Walter Baade. "I have invented something which can never be detected experimentally."

But things were not quite as hopeless as Pauli believed. Although the neutrino interacted less with matter than anything that physicists had previously discovered, detailed calculations showed that once in a great while it was stopped in its tracks. If one could only create a beam of neutrinos that was intense enough, it might be possible to observe neutrino interactions after all.

Every time that a neutrino passes an atom, there is an infinitesimal chance that the neutrino will be captured by one of the particles in the atom's nucleus. If one could set up a barrier consisting of thousands of light years of lead, there would be some hope of stopping a single neutrino. Naturally, conducting such an experiment would be somewhat impractical. But if one has a very large number of neutrinos, say billions of billions of them, things become a little easier. One does not have to stop all of the neutrinos after all; it is only necessary to capture a few.

The neutrino was discovered in 1956 by the American physicists Clyde L. Cowan, Jr., and Frederick Reines. Cowan and Reines set up an experiment next to a nuclear reactor at Savannah River, South Carolina, which gave off an estimated 10^{18} neutrinos each second. For detectors, they used large tanks of water.

Neutrinos cannot be observed directly. But when they interact with matter, particles called *positrons* are given off. The positron is the *antiparticle* of the electron and is relatively easy to detect. It has the same mass as an electron but is positively, rather than negatively, charged. When a positron and an electron encounter one another they are annihilated, and gamma rays are created in their place.

Even though their equipment was able to absorb only one neutrino once every twenty minutes or so, they were able to show that it had to be neutrinos that were responsible for the positrons they saw. The elusive particle which Pauli had thought could never be found had been discovered.

The discovery of the neutrino was an impressive experimental feat, but it was a triumph for theoretical physics. Until Cowan and Reines performed their experiment, it was possible to believe that physicists simply did not know what they were talking about when they discussed beta decay. But now a particle whose existence had been deduced on purely theoretical grounds had turned out to be as real as the proton, the neutron, and the electron. It was true that most of the neutrinos which were created during the decay of radioactive atoms never interacted with matter again. But a few did, and the detection of some of those few demonstrated that Pauli's particle was real.

According to modern theory, there are relatively few fundamental particles in nature. It is well established that particles such as neutrons and protons are made up of even more fundamental particles, called *quarks*. So are the numerous other baryons, and the mesons as well. A small number of particles do not have quark constituents. These are called *leptons.*

The most familiar lepton is the electron. The other particles in the family are the *muon* and the *tau* particle (the names come from the Greek letters *mu* and *tau* which are used to designate them), and several different kinds of neutrinos. In recent years it has become apparent that there is more than one variety of neutrino. Physicists think that there are probably three, one corresponding to the electron, one that is paired with the muon, and one for the tau lepton.

The muon is a particle which resembles the electron. The only significant difference between the two seems to be the fact that the muon is 207 times as heavy. No one knows why the muon should exist; it seems to perform no essential role in nature. There are not even very many of them around at any given moment: Once created, a muon will decay into an electron in about a millionth of a second.

The discovery that there was more than one kind of electron-like particle caused physicists to wonder whether there might not also be more than one kind of neutrino. After all, the fact that the electron and the neutrino divided up energy with one another in the beta decay process showed that they were intimately related. Couldn't there be other kinds of neutrinos that were related to the muon and the tau lepton?

This suspicion turned out to be correct. In 1962, Columbia University physicists Leon Lederman and Melvin Schwartz performed an experiment at the Brookhaven National Laboratory in which they showed that there are indeed both an *electron neutrino* and a *muon neutrino*. They showed that neutrinos created during the formation of muons could produce other muons when interacting with matter, but never ordinary electrons or positrons. The electron neutrinos, on the other hand, never become involved in processes which lead to the creation of muons.

No one was sure what the difference between the electron neutrinos and the muon neutrinos was. The two varieties were alike, it was thought, in that they had zero charge and zero mass, and the same amount of spin. And yet the subatomic particles with which they reacted could apparently tell one kind from the other.

For every particle that has been discovered, there exists an antiparticle. Physicists do not really know why this should be the case. It is one of the symmetries that is exhibited by nature, however, and must be accepted as a fact. When a particle and its antiparticle collide, they mutually annihilate each other, disappearing in a burst of radiation. Physicists think that the particles in the universe far outnumber the antiparticles. If this were not the case, *antimatter*—matter made of antiparticles—would be common in the universe, and we would see violent explosions whenever matter and antimatter encountered one another.

The neutrinos provide no exception to this rule. If there is an electron neutrino, there must also be an electron antineutrino. If a muon neutrino exists, so does the muon antineutrino. In fact, it is actually an antineutrino that is given off in the beta decay process.

If the muon can be thought of as a heavy electron, then the tau lepton, which was discovered in 1977, is a "heavy-heavy electron." Its weight is about 3,500 electron masses. Although no tau neutrino has yet been found, scientists are confident that it will be discovered. If the tau did not have a neutrino, the symmetry that nature has exhibited so far would be violated. Like the muon and electron neutrinos, the tau neutrino was thought to lack charge and to be massless. However, it very soon became apparent that contrary to what physicists had always believed, neutrinos might have mass after all.

The units in which the masses of particles are customarily given are not pounds, or kilograms, or even milligrams. Subatomic particles are so small that measuring them in such units would be very impractical. The unit of mass that is most often used is the *electron volt,* which is abbreviated as "eV." Now, strictly speaking, the electron volt is a unit of energy, not mass. However, the fact that mass and energy are equivalent makes it possible to use the electron volt to measure either. The proton weighs 938.2 million electron volts (938.2 MeV for short), the neutron 939.6 MeV, and the electron 0.511 MeV (which is easier to write than 511,000 eV). By comparison, the upper limit that scientists had obtained for the neutrino mass was very small, only about 60 eV.

In the spring of 1980, three University of California at Irvine physicists, Frederick Reines (the same Reines who back in 1956 had proved, with his collaborator Cowan, that the neutrino exists), Henry W. Sobel, and Elaine Pasierb, announced that they had discovered that the neutrino did have a nonzero mass.

Reines, Sobel, and Pasierb did not try to measure neutrino mass directly. They attempted to see, instead, whether neutrinos oscillated from one variety to another. In 1968, Bruno Pontecorvo had showed that it might be possible for one kind of neutrino to spontaneously transform itself into another. For example, an electron neutrino might suddenly turn into a muon or a tau neutrino. After a period of time, the tau or muon neutrino would turn into an electron neutrino again.

According to currently accepted theories, neutrinos can oscillate only if they have mass. If the oscillation is found, then the mass is not zero.

Reines, Sobel, and Pasierb thought they had detected oscillations between the electron and tau neutrinos. The muon neutrino was apparently not involved. The experiment contained a lot of uncertainties, not the least of which lay in the fact that the tau neutrino was not detected (when some of the electron neutrinos disappeared for a while, it was assumed that they had changed themselves into the tau variety). Because of the possible errors, even Reines did not believe that the results were conclusive. He agreed that other experimenters would have to confirm the results before it would be possible to say that neutrino oscillations had definitely been found.

When similar experiments were done elsewhere, the results were negative. The other experimenters could not succeed in detecting neutrino oscillations.

When experimental results cannot be confirmed, it is generally assumed that they are spurious. The experimental apparatus that scientists use nowadays is very complicated. However careful one is, unknown sources of error do creep in. If it appears that these errors might be significant, other physicists will try to repeat the experiment under different conditions. Although the second experiment may contain sources of error too, it is likely that they will be different ones. In any case, the second experimenter will usually be able to improve the design of the experiment in one way or another, so that the accidental error is smaller.

In this case, attempts at confirmation failed. It is likely, therefore, that the Irvine physicists' results would have been dismissed and forgotten if it were not for the fact that at about the same time, Soviet physicists reported that they too had detected neutrino mass.

The Russian experiment, carried out by a group of scientists at the Institute for Theoretical and Experimental Physics in Moscow, had nothing to do with neutrino oscillations. The Russians attempted to measure the neutrino mass directly. Reporting on their results they claimed to have found—with a certainty of 99 percent—that the mass of the neutrino was somewhere between 14 and 48 eV. Their experiment said nothing about the masses of the muon and tau neutrinos; they had attempted only to weigh the electron neutrino, the most common variety. However, if one kind of neutrino has mass, it is only reasonable to assume that the other kinds have mass also. The masses will not necessarily be the same, but should not be zero.

The Russian scientists made very accurate measurements of the energy of the electrons emitted in a certain beta decay process. They felt that if they could accurately determine the energies of the electrons, they could calculate both the energy and the mass of the neutrinos that were simultaneously given off. For this purpose, they used tritium, or hydrogen 3. Tritium releases very-low-energy electrons when it undergoes beta decay, making the difficult task of measuring neutrino mass a little easier.

But the Russian experiment did not settle the controversy. If anything, it caused the level of scientific argument to increase. Some scientists thought that there were hidden errors in the Soviet experiment. They pointed out that the measurement had been a very difficult one and that, as a result, the outcome of the experiment was not as conclusive as it seemed.

At the moment, scientists working on the problem agree that it is necessary to do much more work, and that years may pass before any definite conclusion can be reached. However, opinion is shifting toward the conclusion that neutrino mass is very likely. While the experimental physicists argue about the proper interpretation of the experiments that have been performed, the theoretical astrophysicists point out that if neutrinos do oscillate or have mass, then a number of outstanding problems can be cleared up.

One of these involves the number of neutrinos produced by the sun. For years Raymond Davis, Jr., a physicist from Brookhaven National Laboratory, has been performing experiments designed to measure the number of neutrinos traveling from the sun to the earth. His results seem to indicate that there are less than half as many neutrinos as there should be.

But if the neutrinos oscillate from one variety to another, the discrepancy between theory and experiment disappears. Davis's experimental apparatus detects only electron neutrinos, the kind that the sun is believed to produce in the greatest numbers. If some of these changed into neutrinos of the muon or tau variety during the 93-million-mile journey from the sun to the earth, then fewer neutrinos would be seen than theory predicts.

And if neutrinos have mass, this could explain the dark haloes which surround galaxies. Massive neutrinos would not travel at the speed of light, as those of the massless variety are presumed to do. They could easily be slowed to velocities of a few kilometers per second, and they could be captured by the gravity of galactic clusters. The dark haloes could therefore very well be made up of slow neutrinos orbiting the visible central regions of galaxies.

According to physicists David Schramm of the University of Chicago and Gary Steigman of the Bartol Research Foundation, neutrinos can account for galactic haloes if their masses are somewhere between 3 and 20 eV. If they are lighter they will never coalesce around galaxies; the gravitational attraction between galaxies and neutrinos would be so weak that the latter would simply zoom by. And if the neutrinos' mass were above 20 eV, there would be more mass around galaxies than is observed.

If Schramm and Steigman's theory is correct, most of the mass of the universe exists in the form of neutrinos. They estimate that if neutrinos weigh more than 1.4 eV, their combined weight is greater than that of ordinary matter. And if the mass is between 3 and 20 eV, then they contribute something between 70 and 95 percent of the mass in the universe.

Neutrinos could also clear up some of the problems concerning galaxy formation. Scientists are fairly certain that galaxies began to form within a few hundred years after the big bang, but have never been quite sure about the details of the process. Furthermore, no theory adequately explains why galaxies are the size they are, or why they should come in clusters.

In 1980 the Hungarian physicist Alexander S. Szalay suggested a possible solution. If neutrinos do have mass, they will attract one another gravitationally. It is therefore possible that shortly after the big bang, gravity caused neu-

THE FATE OF THE UNIVERSE

trinos to cluster together. The more the neutrinos coalesced, the greater the gravitational attraction became. In many respects the process would resemble the contraction of clouds of gas that evolve into stars.

Gradually the universe cooled down. After several hundred thousand years, ordinary matter began to accumulate in the clumps of neutrinos. Once galaxy formation started, matter accumulated very quickly. According to this theory, there is no reason to invoke any of the numerous *ad hoc* assumptions that have been used to explain galaxy formation.

If Szalay's theory is correct, the neutrinos that fill the universe today already existed shortly after the big bang— an assumption that is rarely questioned. Even before it was suggested that neutrinos might have mass, physicists generally agreed that most neutrinos were created out of the radiation that filled the primeval fireball. The numbers that are created in beta decay processes are so few as to be utterly insignificant by comparison.

Szalay speaks of galaxies which form inside clumps of neutrinos, while Schramm and Steigman picture already-existing galaxies which capture neutrinos as the latter fly through space. The two theories may seem contradictory, but they really are not. Schramm and Steigman's theory does not depend on the assumption that galaxies came first; the important point is that the dark haloes can be explained by neutrino mass in a certain range.

According to Floyd Stecker of the NASA Goddard Space Flight Center in Greenbelt, Maryland, neutrinos with mass in this range may already have been observed. Stecker points out that if there exist light and heavy neutrinos, then radiation will be given off when one kind decays into the other. If the heavy neutrino has a mass of around 14 eV, this decay should produce a slight glow of ultraviolet radiation in the sky.

Stecker points out that exactly such a glow has been observed by a group of physicists at Johns Hopkins University, and also by a group of French scientists working with the French D2-B spacecraft. Observation of this glow does not demonstrate conclusively that 14 eV neutrinos exist; the ultraviolet radiation could ultimately turn out to have been caused by something else. However, the fact that Stecker's theory should imply a mass of this size is interesting, to say the least, for 14 eV is practically in the middle of the 3-to-20-eV range of the Schramm-Steigman theory. Moreover, the figure agrees reasonably well with the experiments that have been performed, controversial as these may be.

As this chapter is being written, the case for neutrino mass is considered quite tentative. However, the evidence does seem to be mounting. The idea is made especially attractive by the fact that it seems to explain so many different things. At the present time, it is this fact—even more than the experimental evidence—which makes neutrino mass seem a likely possibility.

But could neutrinos close the universe? David Schramm thinks not. A neutrino mass between 3 and 20 eV would not be quite sufficient. Furthermore, the neutrinos would be concentrated in galactic haloes. And, as we have seen, the haloes alone could not provide the missing mass.

However, the arguments in favor of an open universe do not seem as strong as they were just a few years ago—especially since no one knows what the masses of the various different kinds of neutrinos are (assuming, of course, that neutrino mass turns out to be a reality), or how they may be distributed throughout the universe. At this point, all one can do is make guesses.

Is the universe open or closed? Scientific opinion no longer favors one possibility or the other. All the arguments have broken down, and every attempt to find out has led to uncertainty. Arguments based on the deceleration parameter floundered when it was realized that the brightness of galaxies probably did not remain constant. The helium-deuterium arguments apply only to ordinary matter, not to neutrinos. The missing mass has not been found, but the discovery of dark haloes demonstrated that the universe contained quite a bit of mass that astronomers had not realized was there.

Is the universe open or closed? Only one thing seems certain. Whether it is one or the other, it is very close to the borderline. It is so close that fifty years of experiments have led to no definite conclusion. It is almost as though the universe had been deliberately constructed in such a way as to provide us with a puzzle that could not be solved.

Naturally I don't intend that the above statement be taken literally. It would be absurd to suggest that a God —or a demiurge—had consciously designed the universe with the intention of playing a practical joke upon twentieth-century scientists. In any case, it has been quite a long time since anyone seriously believed that scientific questions could be answered by trying to understand the motives of a deity. But I do suggest that there is something very puzzling about the universe.

In particular, why should it be so close to the borderline? Is there some good scientific reason why the universe should have been put together in such a way as to make it very difficult to tell whether it is finite or infinite? Is this just a cosmic accident? Or is there some hidden logic behind it all?

Although these questions sound perhaps more metaphysical than scientific, attempts have been made to answer them. This is the subject that I will discuss next—after I say a few words about how the universe is destined to end.

CHAPTER 7

The Fate of the Universe

AS we have seen, stars contribute only a small fraction of the mass of the universe. Far more exists in the form of interstellar gas, and possibly in neutrinos. However, stars are certainly the universe's most prominent feature. To discuss the fate of the universe without mentioning them would be like talking about the ecology of a forest without making any mention of the trees.

Stars come in a variety of sizes. The smallest have only a few percent as much mass as the sun, while the most massive are many times larger. The mass of a star is directly related to its life span. The largest stars—the bright, hot *blue giants*—are the ones which burn out first; they gobble up their nuclear fuel at the most rapid rate. For example, a star with thirty times as much mass as the sun has a life span of only a few million years. Very small stars, on the other hand, can go on shining for hundreds of billions. The sun, whose mass lies approximately halfway between the two extremes, has been shining for about 4.5 billion years, and is expected to continue doing so for about 5 billion years more.

There is much that is not yet understood about the later stages of stellar evolution. In particular, astronomers are not sure that they have worked out all the details of a supernova explosion. However, one thing is clear. Most, if not all, of the larger stars end their lives as supernovae, while the smaller ones die in a more peaceful manner. After spending brief periods—a few hundreds of millions of years—as red giants, the latter shrink into white dwarfs. As the white dwarfs cool, their light output gradually decreases, and they evolve into black dwarfs. Or at least that's the way it should work. The universe seems not to have existed long enough for any stars to evolve into black dwarfs yet. Consequently the only black dwarfs that exist right now are bodies that were not quite big enough to evolve into stars in the first place.

The appearance of the galaxies will not change much in the next few billion years. Some stars will die, and will be transformed into white dwarfs, neutron stars, and black holes. But new stars will be born to take their place. The interstellar gas from which they are formed still exists in generous supply. At the present time, only a few percent of the available nuclear fuel in the universe has been used up. As a result, we can expect that there will be stars in the sky for quite a long time.

But nothing goes on forever. It is estimated that 100 billion years from now, the galaxies will be dying. At that time only a very few dim stars and white dwarfs will still be shining. The supplies of interstellar gas will be almost entirely exhausted, and galaxies will primarily consist of stellar corpses.

If the universe is open, the expansion will go on forever. As a result, even as galaxies die they will continue to recede from one another, and there will come a time when other galaxies are no longer visible, even to the most sophisticated astronomical devices. If intelligent beings still exist in the Milky Way at this point, the Milky Way will appear to them to be the only galaxy that exists.

As the universe inexorably expands, other events will be taking place. The densely packed stars in the galactic cores will coalesce into black holes with masses hundreds of millions, or possibly billions, times greater than that of the sun. Indeed, there is evidence that something like this might already be happening. Astronomers have observed numerous galaxies that are producing very large quantities of energy in their cores. The most plausible explanation seems to be that these cores contain supermassive black holes, and that energy is released when interstellar gas falls into them. Of course if such black holes are forming 100 billion years in the future, there won't be much energy production. By that time, most of the instellar gas will have been used up, if it has not evaporated into intergalactic space.

Over periods of tens of billions of years, the black holes that inhabit the centers of galaxies will grow larger as they gobble up more and more matter. Random collisions will send stars spiraling into the black holes now and then. As these are absorbed, the holes will grow larger, and stars will be gobbled up at an ever more rapid rate. Eventually they will absorb all of the matter that is presently seen in galaxies, except for those stars that "evaporate" into intergalactic space after suffering stellar collisions that thrust them in just the right direction.

The growth of the black holes will continue even after the galaxies are gone. Entire clusters of galaxy-sized black holes will coalesce, and most of the stars that had previously avoided absorption into the black holes will now be gobbled up. After a period of time that has been variously estimated as 10^{19} or 10^{29} or 10^{34} years, the universe will be made up primarily of galactic and supergalactic black holes, with a few black dwarfs, neutron stars, stellar black holes, and planets in the spaces between them.

After about 10^{100} years the galactic black holes will evaporate. The process is the same as the one which has been discussed in connection with mini black holes. To describe it in detail would lead us too far astray, but the evaporation is a consequence of the laws of quantum mechanics. These laws say that every black hole must suffer such a fate, and that the amount of time that must elapse before it happens is related to the black hole's mass. Since the black holes that will exist at this time will have hundreds of billions or trillions of solar masses, it is obvious that the time must be long indeed.

If current theories of elementary particle interactions are correct, the matter that is not contained in black holes will already have disintegrated. It is now believed that the proton, one of the basic constituents of atomic nuclei, has a life span of about 10^{40} years. This fact has not yet been confirmed experimentally; but if it is true, matter will not last forever. Atoms cannot exist without nuclei, and nuclei cannot exist without protons.

When the black holes evaporate, numerous particles and

antiparticles will be created. Most of these will annihilate one another, producing radiation. Those that do not will be caught up in a universe that is remorselessly expanding. If any matter still exists at this point, it will be ever more thinly dispersed.

As the expansion goes on, radiation will also become weaker and weaker. Eventually the contents of the universe will be so diffuse that, for all practical purposes, there will be nothing left but an empty but still-expanding space.

Many people—scientists and lay persons alike—find the prospect of a universe that will evaporate into nothing a rather depressing one. However, the fate of a closed universe is not much better; a closed universe must eventually collapse and be destroyed in a fireball similar to the one in which it was created.

In a closed universe, the expansion must eventually stop. It is impossible to say how much time must elapse before this happens; it depends upon how quickly the expansion is slowing down. The best we can do is to place a rough lower limit on the amount of time required. Since the universe is still expanding very rapidly, the minimum time must be something on the order of tens of billions of years, and it is perfectly conceivable that trillions of years could go by before the universe slows down and stops.

Once the expansion does halt, gravity will begin to pull everything together again. If at this time the galaxies are extremely dispersed, the contraction will be very slow at first. But since nothing will counteract the gravitational attraction, the universe will contract. The closer together the galaxies come, the greater their attraction will be. As a result, the contraction will proceed at an ever increasing rate.

If stars and galaxies still exist when the contraction

starts, the universe may not look much different than it does today. The galaxies will be farther apart, and the stars will be dimmer; it is possible that a few young, bright galaxies may have formed out of the gas in intergalactic space. The difference will be that as the contraction begins, the redshifts that we observe today will become blueshifts.

As the galaxies come closer together, they will begin to collide and merge with one another. However, it will not be an especially catastrophic event. The stars within galaxies are so far apart that stellar collisions will be very rare. The chance that any given pair of stars will crash into one another is less than the probability that two bullets fired by soldiers who are shooting at one another will collide in mid-flight.

Billions of years will pass, and the galaxies will be compressed into a space that is growing smaller with every passing moment. The dark, empty spaces that make up most of the universe today will disappear. The night sky will be packed solid with stars. Galaxies, including our own, may contain more stars than they do now. As the universe becomes smaller, gas will become more compressed. As a result, more material will be available for star formation.

If new stars are formed in this manner, most of them will not last long. Some of them will be destroyed by collisions; as the universe becomes denser, this will happen with increasing frequency. Those stars that survive collision will be torn apart by the intense radiation that will soon fill all of space.

Some of this radiation will have its origin in starlight. Since starlight cannot leave a closed universe, it will accumulate. As the universe contracts, the starlight will become more intense and increase in frequency. Visible light will be transformed into ultraviolet radiation, and then into x rays and gamma rays.

The same thing will happen to the microwave radiation background. These radio waves, which today have a radiation temperature of only three degrees above absolute zero, will gain in energy. Eventually they will make an even greater contribution to the radiation density than the starlight does.

As the radiation vaporizes the stars, the latter will take the appearance of streaks in the direction that they were moving. And then as the temperature rises even higher, the streaks will disappear; the hot gas that now fills the universe will dissipate them.

During all this time, the contraction will have been proceeding at an ever faster rate. The temperatures will rise higher and radiation will become more intense. Eventually even atomic nuclei will be destroyed. The universe will condense into a "soup" of radiation, subatomic particles, and black holes.

Even though there is not yet any incontrovertible proof that black holes exist, most astronomers have become convinced that they must be forming in great numbers right now. If they are, then they should have become many times more numerous by the time the universe begins to contract. Like the black holes that will form in an open universe, some of them may have gobbled up entire galaxies by then.

As the universe collapses, the black holes will grow. As they do, they will absorb matter and radiation at an increasing rate, and grow larger still. Then, as the contraction brings them closer together, they will begin to coalesce. Before the collapse comes to an end, some of them may have masses millions or billions of times greater than that of our galaxy. It is perfectly reasonable to think of such black holes as portions of the universe that have collapsed already, as though they were not willing to wait for the end of the contraction.

As the universe is squeezed into an ever decreasing volume, the matter and radiation that have not disappeared into black holes will be squeezed out of existence. As they are, the black holes will merge with them, and the entire universe will be compressed into a singularity. This process, the reverse of the explosion in which the universe began, is sometimes referred to as the "big squeeze" or the "big crunch." If the big bang was the beginning of the universe, then the big crunch will be the end.

Some scientists have found the idea that the universe may cease to exist rather hard to swallow. And, to their way of thinking, the idea that there was a creation—the big bang—at another point in time is even worse. They feel that there are philosophical reasons for hypothesizing that the universe has always existed.

Speculation that the universe might have existed in some form before the big bang led to the *oscillating universe* theory. According to this theory, which was proposed by Friedmann in 1922 and expounded in detail by the American physicist Richard C. Tolman ten years later, the universe goes through one cycle of expansion and contraction after another. It is not destroyed in the big crunch. Instead, it somehow "bounces" and re-explodes in a new big bang.

This idea seems very appealing. Physicists had found themselves uneasy with the idea of a universe that burst into existence. Although there seemed to be no good scientific objections to the idea of creation out of nothing, it couldn't easily be incorporated into existing theory. *Ex nihilo nihil fit* goes the Latin phrase. Nothing is made out of nothing. This may be only metaphysical prejudice, but it has led to quite a bit of speculation about the universe before the big bang, and about how it might evolve after the big crunch.

Offhand, one might think that nothing could be more straightforward. The universe expands, collapses, and expands again *ad infinitum*. One does not have to worry about where the universe came from; it was always there. And it is not necessary to contemplate a future in which all life will cease to exist, because life will be created anew in each cycle.

However, it has become apparent in recent years that certain theoretical difficulties are associated with the oscillating universe idea. One has to do with the fact that a bouncing universe resembles a bouncing ball in that each cycle should be a little different from the previous one. But unlike a ball, the universe bounces "higher" and higher each time. That is, in each successive cycle it expands for a longer time before it begins to contract. The cause of this effect is the accumulation of starlight. Radiation that is present in one cycle should be carried over to the next. The starlight that was emitted in every previous cycle of the universe should fill the space around us; it is the additional radiation density that causes the increase in the size of the bounce.

If the cycles of expansion and contraction will grow longer in the future, then they must have been shorter in the past. It is possible to calculate that in this case there can have been at most a hundred bounces. In other words, the universe must have had a beginning at a finite time in the past. We are thus led back to the "problem" of creation out of nothing that the oscillating universe was designed to avoid in the first place.

But there are ways of avoiding such a conclusion. One need only make additional assumptions. The resulting theories often seem bizarre, but to say this is not to criticize them. The universe knows nothing of human ideas of what is "bizarre" and what is not. Indeed, it could very well turn out to be stranger than anything that we can imagine.

The English physicist Thomas Gold evades the difficulty associated with lengthening cycles by assuming that when the universe reaches its point of maximum expansion, time begins to flow backward. As it does, starlight flows back into the stars, black holes disappear and ordinary stars appear in their place, and the matter that has been ejected into space by supernova explosions comes back together again.

If the earth no longer exists when the expansion stops, it is recreated. On this new earth, everything happens in reverse. Rivers run uphill, rain rises from the earth to the sky, and energy flows back into the sun. People grow younger, not older; their lives begin when they rise up out of their graves, and end when they enter their mother's womb to become embryos.

It might seem that even though we agreed that some strange theories would have to be considered, this one is simply too bizarre to deserve serious consideration. However, this is not really the case. In Gold's reversing universe, people's mental processes would run backward too. Since sequences of events and their perceptions of those events would both be reversed, the universe would have exactly the same appearance as one in which everything ran in a "forward" direction.

If Gold's theory is correct, we have no way of telling whether we are living in a "normal" or a time-reversed phase. There is no way to distinguish between the two. It is not even possible to say that time runs "forward" in one and "backward" in the other. Those two terms have ceased to have any real meaning; choosing one direction of time as the "forward" one becomes a matter of convention.

In Gold's theory, the universe has neither a beginning nor an end. It oscillates back and forth between the big bang and the big crunch. What we view as the collapse of the universe would be seen by the people in the other half as the big bang. In Gold's universe, time is cyclical.

Naturally there are a lot of paradoxes associated with such a theory. Suppose, for example, that the expansion of the universe is not perfectly uniform, and that it begins to contract in certain regions while the expansion continues for a while elsewhere. Will time then flow in both directions simultaneously? Or suppose that a creature seals itself in a vault that makes it invulnerable to all outside influences. Would it not continue to experience a time that continues to flow in the same direction while the rest of the universe reverses? And would it not be able to come out of the vault and interfere with causality in the region of the reversing universe that it now occupies?

And how are black holes to fit into such a theory? It is possible to think of black holes as small regions of space that have already collapsed, long before the big crunch. Are we to assume, therefore, that time runs backward in black holes? Or do the black holes wait until the expansion of the universe ceases before they reverse their time flow?

These problems must be dealt with before Gold's theory can be taken seriously. If a theory is to be judged correct, it must not contain paradoxes that allow one to make contradictory predictions.

There is a way to perform an experiment that would test Gold's theory. If it is correct, there is no reason why messages could not be sent from one half of the universe to the other and it is possible for us to look to see whether any such messages are being received.

Of course they would not look like ordinary signals. Since they would have been sent from the half of the universe in which time runs in the opposite direction, they should show up as power drains on radio transmitters beamed into space.

In 1973 the American astronomer R. B. Partridge performed an experiment designed to determine whether such power drains could be observed. Though the experiment was not successful, it does not invalidate the theory. It is possible that no messages from the future have been sent. Or perhaps more sensitive experimental apparatus will detect them in the future. However, even though the experiment cannot be considered conclusive, most scientists do not take Gold's theory seriously.

The paradoxes associated with the theory can be avoided if one is willing to make certain modifications. Just such a modification has been proposed by the English cosmologist and mathematician Paul Davies. Davies suggests that time reverses its direction not at the moment of maximum expansion, but in the big crunch. After the crunch, the universe goes through a second expansion-contraction cycle in which time is reversed.

The only difference between Davies' theory and Gold's is that in the former, the universe has two cycles instead of one. Both theories make the assumption that time will loop back upon itself. In either case the universe has no end, only two beginnings.

In Davies' theory the paradoxes are avoided because the "normal" and "time-reversed" halves of the universe are separated by a natural boundary, the big crunch (it is seen as a crunch from both sides of the universe, by the way) through which nothing can pass intact. This eliminates the possibility that there might ever be a region of space where time is flowing both ways at once. Unfortunately, it also eliminates the possibility of performing an experiment such as the one described above. If Davies' theory is correct, we cannot detect messages from the future; they cannot pass through the big crunch intact either.

The ideas of reversed and cyclical time are not modern ones. The reversed-time theory is expounded in Plato's dialogue "The Statesman." A "stranger" explains to a somewhat credulous young Socrates that it is the Creator who causes time to flow in a given direction. As soon as He lets go, it will reverse. It is only natural that it should do so; after all, the Creator has been forcing it a certain way for so long a time. When He finally does let go, the stranger says, time will first come to a standstill, then people and animals alike will begin to grow younger.

Some of the ancient adherents of the Stoic philosophy (which originated in Greece and became quite popular among the Romans) describe a kind of cyclical time which is very similar to that encountered in the theories of Davies and Gold. According to the Stoics, the world was destined to be destroyed by fire, and to be recreated in the same form. In the next cycle of the universe, the same events would take place all over again.

Today there is no more reason to take such theories seriously than there was in the time of Plato or of the Stoics. On the other hand, there is no way to disprove them. So even if we do not find ourselves convinced that such things can happen, we must still admit that modern cosmological speculation has enlarged the realm of the possible, and has showed that some of the ideas which we encounter from time to time when reading the ancients are not as fantastic as we had thought. At the very least, Gold's and Davies' theories show that we cannot blindly make the assumption that time must always flow in the same direction.

Time-reversal theories are not the only way to avoid the conclusion that an oscillating universe must have cycles that grow progressively longer. If the accumulated starlight is destroyed or transformed into matter in the big crunch, the slate would be wiped clean, and the new universe emerging from the subsequent big bang would contain nothing that bears the traces of previous universes.

This assumption is made in the *superspace* theory of University of Texas physicist John Archibald Wheeler. In fact, Wheeler goes even further, making the additional assumption that the very laws of nature might change when the universe collapses and re-explodes. In Wheeler's view, the universe is "reprocessed" as it bounces so that it is born anew with new laws and physical constants. In the new universe, gravitation may work in a different way, and constants such as the speed of light and the charge on an electron might very well be different.

Wheeler's superspace has little to do with ordinary

space. On the contrary, it is an abstract mathematical entity with an infinite number of dimensions. At any given point in time, the universe occupies a single point in superspace. The theory does not imply that we live in an infinite-dimensional universe, it is just a precise mathematical way of describing the different possible configurations of the universe. Unlike the dimensions of real physical space, these may very well be infinite in number.

Wheeler's theory is a very speculative one. Even though there does not seem to be any way to test it (it would be necessary to travel into the next universe to see if physical laws had changed), the very fact that a superspace theory is conceivable should cause us to question certain of the assumptions upon which the oscillating-universe theory is based.

The latter says that cycles will become longer because starlight accumulates from one cycle to the next. But how do we know that it does? During the collapse of the universe, densities will become so great that all the known laws of physics will break down.* If that is the case, how can we possibly predict what will happen to the accumulated starlight? And if we can't tell that, how can we be sure how it will affect the bounces?

British physicists Roger Penrose and Stephen Hawking have proved a series of theorems which seem to indicate that if general relativity is correct, then a singularity in the collapsing universe (or in black holes) cannot be avoided. There is no way that the infalling pieces of matter can

---

* This is not the same as saying that they will change. There are presumably unvarying laws which describe the behavior of matter at extremely high densities; we simply don't know what they are.

"miss" one another and fly outward again; everything must be crushed into a point of infinite density.

As I have pointed out before, it is perfectly possible that the density of matter does not become infinite; it is expected that general relativity will break down before that occurs. However, there are good reasons for believing that at the very least, the density of matter will grow almost unbelievably large.

It is thought that general relativity should remain valid down to regions that are smaller than an atomic nucleus. If it does, it may be that the entire universe will be crushed into a volume as small as or smaller than this before any unknown effects can halt its collapse. And if the universe does experience this massive a crunch, it is hard to believe that any of the physics that we know will still be valid.

This implies, among other things, that we are free to imagine almost anything we like. We are free to believe, as the Stoics did, that the universe repeats itself endlessly. Like Plato's stranger, we can believe in the idea of reversed time. And of course it is always possible to think that time and space will simply come to an end in the big crunch.

Even if it turns out that our universe is open, this fact need not put a damper on speculation. According to Princeton University physicists Robert H. Dicke and P. J. E. Peebles, it is conceivable that universes can reproduce themselves. In such a case, even a universe that expands forever could produce offspring that go on oscillating forever (or at least long enough to produce offspring of their own). In such a case, universes could still exist in infinite variety.

Dicke and Peebles point out that there is no reason why

an oscillating universe could not become more massive in each successive cycle. Although they do not say how this might happen, they ask us to imagine that a universe is "fattened" at each bounce.

If this could happen to a universe, the same process could presumably take place in a black hole, which is just a collapsing universe in miniature. The black holes in an open universe could experience mini big crunches (or "minicrunches" if you prefer) and rebound in mini big bangs. Naturally we would not be able to see these things happen; these are events that would take place within the event horizon. We would never be able to detect what was going on as long as black holes did not grow into new universes too rapidly.

In the theory of Dicke and Peebles, everything would not come to an end when an open universe evaporated into nothing. By this time, the black hole "seeds" might well have grown into universes themselves. The objection that all black holes must evaporate sooner or later does not contradict this idea, because evaporation need not take place if the black holes are constantly increasing in size.

It is conceivable, therefore, that our universe already contains countless "mini universes" that will eventually grow to the size of our own. If they do, these universes will be forever isolated from one another. It is impossible to escape from a black hole, even if the universe that originally surrounded it no longer exists.

It is not even necessary that stars be formed before black hole seeds can be created. Dicke and Peebles' theory works just as well with mini black holes. Furthermore, such mini black holes could be created out of nothing—creation *ex nihilo* with a vengeance.

144

The Fate of the Universe

Dicke and Peebles imagine a situation in which a few particles come into existence in a kind of random fluctuation. According to quantum mechanics, such fluctuations occur all the time. However, since these *virtual particles* are created in particle-antiparticle pairs, they annihilate one another before we notice they are there. In the quantum world, space can never be really "empty." It is full of particles which pop into existence for times on the order of 10^{-24} seconds and then disappear again.

But suppose that a few of these virtual particles get close enough together to form a black hole before they are annihilated. According to conventional theories, this should not make any difference, for the mini black hole that is thereby formed will evaporate at once. If black holes are made in this way, there will be no observable effects.

But in the Dicke-Peebles theory these black holes do not evaporate. Instead, they begin to undergo universe-like oscillations and to increase their mass. Eventually they evolve into full-fledged universes.

If some of the theories we have been considering are bizarre, then this one is absolutely outrageous. But, again, it does not necessarily follow that it cannot be true.

At this point, we must recapitulate and distinguish fact from speculation. When we are discussing what might be true, it is best not to place limits on our imagination. If we fail to understand the universe it may not be because our theories are crazy, but because we lack the imagination to make them crazy enough. But we must not make the mistake of confusing fact with speculation.

The only thing we know with absolute certainty is that the light that reaches us from distant galaxies is redshifted. From this fact, astronomers deduce that the universe is expanding. This is the only explanation that seems to be at all reasonable. There have been other theories, but not any that seem capable of explaining all the observed facts.

We know that there was a big bang; this fact is now considered to be well established. The big bang theory provides us with a very natural explanation for the expansion of the universe. It also explains the cosmic microwave background and the observed amounts of helium and deuterium. Other theories proposed do not adequately explain all the facts, and have been discarded.

Finally, we know that the universe is approximately 10 or 20 billion years old. As we have seen, there is currently some controversy as to the exact age. However, no one doubts that the age has been established to the correct order of magnitude.

These things are really all that we know. We do not know what, if anything, happened before the big bang. We do not know whether the universe is open or closed, whether it will expand forever or come to an end in a fiery collapse. Assuming that it does collapse, we do not know whether or not it will re-explode. We do not know whether successive cycles in an oscillating universe would be very much like ours, or very different. And of course we do not know what goes on inside black holes.

Perhaps it is our lack of knowledge that gives cosmological speculation the fantastic quality it has. As long as we do not know what is going on, we must remain open to anything. If we cannot tell what is true, we must at least try to find out what might be true. If we are to be successful in this, then it is necessary that we stretch our imaginations to the limit.

CHAPTER 8

Other Universes

IN recent years, the idea that our universe must be filled with life has become a scientific commonplace. Most scientists now believe that life must inevitably evolve whenever conditions are suitable. And unless there is something fundamentally wrong with our conception of the universe, suitable conditions exist in quite a few places.

There are 200 billion stars in the Milky Way galaxy, and our galaxy is only one of billions. It is estimated that if the universe is closed, it must contain at least 10 billion galaxies. If the universe is open the number of galaxies is infinite. Therefore, if planets exist around most stars—and, as we shall see, there is every reason to think that they do—then there must be billions of billions of places where conditions are reasonably earthlike.

Scientists are not so chauvinistic as to insist that these

are the only places where life could evolve, or that extraterrestrial organisms must necessarily resemble those found on our planet. They admit that there might be life forms that breathe methane, or that have chemistries based on silicon rather than on carbon. It has even been suggested that life could conceivably exist on the surfaces of neutron stars. However, until we have some evidence that such forms of life are indeed possible, it is best to speculate only about the existence of life that is reasonably earthlike. At least we know that one kind of life is possible. If we find convincing reasons for thinking that it exists in many places, then we can conclude that the universe is teeming with life whether the other kinds exist or not.

During the early part of the twentieth century, it was believed that extraterrestrial life must be quite rare, if it exists at all. At the time it was thought that the planets in our solar system had been formed when the sun suffered a near collision with a passing star. But since the stars in any galaxy are very widely spaced, such occurrences must be extremely rare. The chance that a star like the sun will pass that close to another star during a 5-billion-year period is only about 1 in 10 billion. It seemed to follow that only a handful of stars could have planets. It seemed that if life does exist elsewhere in our galaxy, it would have had a chance to evolve in at most two or three places. There would of course be planets around more stars than that. But some of these stars would be too hot to support life, and some too cold. Some of the planets would be too small to retain atmospheres, and some would be too close to their suns. In other locations, life might begin, only to be destroyed when a star expanded into a red giant. One could, if one wished, assume that life had been created on the earth alone. At the time, there was no scientific evidence to contradict that belief.

By the late 1930s, it had become apparent that there were difficulties connected with the collision hypothesis. Calculations showed that material drawn from the surface of the sun by a passing star would not have condensed into solid bodies; instead, it would have dissipated into interstellar space. The planets, it seemed, should not be here at all.

In 1943, German physicist Carl von Weizsäcker revived a theory originally proposed by Laplace and by the German philosopher Immanuel Kant during the eighteenth century. Weizsäcker improved upon the Laplace-Kant hypothesis by casting it in mathematical form. This allowed detailed calculations that made it possible to test the theory against observational data.

The Weizsäcker theory was a resounding success. It soon became apparent that it not only showed why there should be planets, it also explained why they should have the kinds of orbital motion and chemical composition that were observed.

According to the theory, the planets condensed out of a disk-shaped cloud of dust and gas which formed around the sun during the early stages of the latter's evolution. Today this theory is universally accepted. Although small modifications might yet be made in it, it is not possible to doubt that it is correct in its broad outlines.

There is nothing special about the sun that would lead us to believe that planets should have formed around it but not around other stars. There should be planets around practically every star in the universe. Many of these planets may be too hot or too cold, to small or too large to be hospitable to life. Some of them may not have atmospheres, and others may orbit hot, massive stars that burn themselves out before the evolution of life has a chance to begin. But even if 99 stars out of 100 were eliminated as possible habitats for life, countless billions of possible sites would remain.

As a matter of fact, 99 percent of the stars cannot be eliminated. It has been estimated that there is 1 reasonably earthlike planet for every 5 stars in our galaxy. This implies that there must be something like 40 billion possible sites for life in the Milky Way alone.

Biologists now believe that life must inevitably evolve whenever the proper conditions exist. All that is needed is an atmosphere like the one that existed on the primeval earth. Provided that there is some source of energy, such as ultraviolet radiation from a star, complex organic chemicals will form spontaneously. Among these are the nucleic acids and amino acids that are found in every living cell. No one has as yet duplicated the steps by which these chemicals presumably came together to make the first living organism. But scientists are convinced that with billions of years available, this process must inevitably happen wherever favorable conditions exist.

It is interesting to speculate about life elsewhere in the universe, especially about the possibility that there might be numerous intelligent species and possibly technological civilizations. However, the really intriguing question may very well be not, "Is there other intelligent life in the universe?" but rather, "Why is the universe so hospitable to life in the first place?"

Even if it turns out that there is something terribly wrong with our ideas about the formation of planets, even if the universe turns out not to be teeming with life, this question still has to be answered. Because, after all, *we* exist. And, as we shall see, this is a fact that has to be explained.

Offhand, one would not think that the existence of terrestrial life and of human intelligence are facts that need to be accounted for at all. "Can't one simply accept the fact that we do exist, and leave it at that?" one is tempted to ask. And, in any case, hasn't it already been pointed out that scientists believe that life is inevitable, given the right conditions? Given the fact that we do have reasonably good conditions on earth, wasn't it inevitable that evolution should have eventually created intelligence?

The answer is, of course, that life and intelligence probably were inevitable. But we must still explain why it is that such ideal conditions existed. It is possible to imagine an infinite number of different kinds of universes. In the vast majority of these universes, life could not possibly arise. In order to be hospitable to life, the universe must be very special. The question that we are really asking is, "*Why* is the universe so special?"

One of the most astonishing things about the universe is its size. The nearest star is a little more than 4 light years away, or about 25 trillion miles. And yet, by astronomical standards, this is a very short distance. Astronomers have observed galaxies that are billions of light years away. At the very least, the universe extends for tens of billions of light years in every direction. And of course if it is open, it goes on forever.

There is another way of looking at this state of affairs: The universe is very spread out. Although stars and planets are relatively dense, there are very great spaces between them. As a result, the universe contains, on the average, less than one atom of ordinary matter for every cubic meter of space. Air is 10^{27} times more dense. And yet we speak of "thin air." Why should matter be so thinly dispersed?

Surprisingly, this question can be answered fairly easily. A universe must be spread out like this if it is ever to give rise to life. In a denser universe, the expansion would halt much too quickly, and there would not be enough time for life to be created.

At the very least, billions of years must pass before life can exist. Galaxies and stars must form. Planets must be created and be given a chance to cool. Nature must have the opportunity to try one chemical experiment after another until the first living organism is formed. A universe that was very much denser than ours would collapse before any of these things happened.

If life is to have the opportunity to exist, a universe must expand out of the primordial fireball at just the right rate. If our universe had been expanding at a rate that was slower by a factor of one part in a million, then the expansion would have stopped when it was only 30,000 years old, when the temperature was still 10,000 degrees. And if the expansion had been faster by a factor of one part in a million, then galaxies could not have formed. Matter would have been flying outward with just enough velocity to prevent it from condensing into clumps.

In other words, the universe must be very close to the borderline between open and closed if life is to have a chance to exist at all. At last we have answered the question, "Why is it so difficult to tell whether the universe is infinite or finite?" If it were possible to tell without much trouble, then there would be no one around to wonder whether the universe was open or closed.

There are yet other ways in which our universe is of a special character. If nuclear forces were just a few percent stronger than they are, there would be no life. Stronger forces would cause all of the primordial hydrogen—not just 25 percent of it—to be synthesized into helium early in the history of the universe. And without hydrogen, the stars could never begin to shine.

As far as we know, there are four fundamental forces in nature: gravity, electromagnetism, and the so-called "strong" and "weak" nuclear forces. Every one of these forces must have just the right strength if there is to be any possibility of life. For example, if electrical forces were much stronger than they are, then no element heavier than hydrogen could form. The positively charged protons would repel one another so strongly that their mutual repulsion could not be overcome by the strong nuclear force. But electrical repulsion cannot be too weak. If it were, protons would combine too easily, and the sun would not burn as slowly and steadily as it does. The protons would combine explosively, and the sun (assuming that it had somehow managed to exist up to now) would explode like a thermonuclear bomb.

If the ratio between the strong and weak nuclear forces were different, the same kinds of things would happen. Either hydrogen nuclei would combine into helium much too readily, or the reaction would simply not take place. We must conclude that small changes in any of the forces of nature would lead to universes in which life would not be possible. Either there would be no atoms, or there would be atoms but no stars or planets. In some conceivable universes, matter would collapse very rapidly into black holes. In others, rapid nuclear reactions would produce cosmic rays of such intensity that biological evolution could never take place.

The theories of modern physics do not tell us why the forces of nature should have exactly the strength that they do, any more than they tell us why the universe should have expanded out of the primeval fireball at just the right rate. For example, the strength of the electromagnetic force (which embraces both electricity and magnetism) is related to a number called the *fine structure constant*. The name comes from the fact that the value of the constant can be determined by studying fine structure in the spectra of light emitted by atoms. The constant has the value 1/137. No one knows why it should be equal to this particular fraction rather than, say, 1/36 or 1/458. However, if it were not very close to 1/137, then life would not exist.

In recent years it has become fashionable to view the creation of life as a stage in cosmic evolution. Galaxies evolved first, and then stars and planets. These produced the conditions necesary for the formation of complex organic molecules and, finally, life. The sequence seems so inevitable that it is difficult to imagine how life could *not* have evolved. As astronomer Carl Sagan states "the origin of life on suitable planets seems written into the chemistry of the universe."

But how did life get written into the chemistry? How is it that common elements such as carbon, nitrogen, and oxygen happened to have just the kind of atomic structure that they needed to combine to make the molecules upon which life depends? It is almost as though the universe had been consciously designed in such a way that life would be inevitable.

Scientists of an earlier age would not have hesitated to conclude that such considerations indicated the existence of a Creator. The German astronomer Johannes Kepler, who discovered the laws of planetary motion upon which Newton's law of gravitation was based, believed that the heavens were an expression of the beauty and harmony of divine creation. Newton concurred, saying that the solar system was "not explicable by mere natural causes," that its structure had to be ascribed to "the counsel or contrivance of a voluntary agent."

The *argument from design*, as this idea is called, is not much in favor nowadays. More than two centuries have passed since Kant in his *Critique of Pure Reason* pointed out flaws in the argument. Although it seems not to have disappeared completely (I recall having heard it in Sunday school as a child), modern theologians no longer depend upon it.

Unlike Newton and Kepler, today's scientists do not believe that there is some region where physics and theology merge with one another. If science uncovers a question, then science should attempt to answer it. But exactly what conclusions should be drawn from the fact that the universe has such a special character? Are we to say that it is all some kind of cosmic accident? That certainly does not sound very satisfactory.

One very obvious way out of the difficulty is to assume that there are an infinite number of universes. The universes that do not have our special character exist but are lifeless. The reason that our universe has certain special properties is that, otherwise, there would be no one here to see it.

It must be emphasized that the hypothesis that universes exist in infinite numbers is anything but accepted scientific theory. However, I do not see how such a conclusion can be avoided. There are simply not any reasonable alternatives.

The idea of an infinite number of universes is not a new one. It is really no more than a modern version of the many-worlds theory of Giordano Bruno and of the Greek philosophers Democritus and Anaximander. The only difference between the modern version and the older ones is that our horizons have expanded somewhat. We speak of "universes" where the Greeks and Bruno talked of "worlds."

The idea that astronomical data implied the existence of an infinite number of universes was first suggested by Robert Dicke in 1961. But Dicke's suggestion did not lead to a great deal of scientific discussion. It may be that it was a little ahead of its time. And when British mathematician Brandon Carter made similar observations around 1968, he did not even publish them at first.

But in 1973 the question was revived by Stephen Hawking and his Cambridge University colleague Barry Collins. Collins and Hawking suggested, in a paper published in *The Astrophysical Journal,* that galaxies—and therefore life—could be created only in a universe that expanded out of the big bang just fast enough to avoid recollapse. The existence of galaxies and of life, they claimed, meant that the universe was exactly on the borderline between open and closed (not approximately on the borderline, as we have previously observed, but exactly).

To Collins and Hawking, this hypothesis seemed to have a certain amount of appeal. The only trouble with it was that the probability that the universe was exactly on the borderline was zero. When a quantity (in this case the expansion velocity) can have an infinite number of different possible values, the chance that it has any one particular value is zero. So Collins and Hawking took the step that had previously been made by Dicke and by Carter. "One possible way out of this difficulty," they said, "is to assume that there is an infinite number of universes with all possible different initial conditions."

The conclusion that there are infinite universes is not the only one that can be drawn. For example, John Archibald Wheeler and American mathematician C. M. Patton have suggested that a universe will only come into existence if it will be able to support intelligent life: There is some unknown factor eliminating all the possible universes that will not harbor intelligent species that can observe them. But the idea that there can be an interaction of this type between observer and universe is a little too mystical for most scientists.

The idea of alternate universes is an old one in science fiction. Anyone who has read much in the genre will certainly have run across short stories and novels in which ideas of this sort are introduced. One of the most common versions is a universe that branches off into near replicas of itself at every moment of time.

This idea is really a very simple one. Nothing is more obvious than the fact that our world is full of chance happenings. There seems to be no good reason why things should happen one way and not another. But suppose, the science fiction authors have said, that the universe splits in two every time one of these chance events takes place. If this is the case, then there will be numerous universes that differ from ours in various large and small ways. In some universes, you never met the person you are married to in this one. In others, you had a son rather than a daughter. In others, the South won the American Civil War. And in yet others, human life never evolved on this planet.

Such stories can be very entertaining. One might not expect to encounter a similar scientific theory, yet such a theory exists. It is called the *many-worlds interpretation* of quantum mechanics, and was proposed by American physicist Hugh Everett III in his Princeton University doctoral dissertation in 1957.

At first Everett's theory did not attract much attention. But as other proposals concerning alternate universes found their way into print, interest in it increased. In recent years it has been championed by Wheeler, among others. It is not consistent with the Wheeler-Patton "self-reference cosmology" described above, but John Wheeler seems to be a physicist who will explore any interesting idea to see where it leads.

Quantum mechanics is a non-deterministic theory. Happenings on the atomic and subatomic levels can be described only in terms of probabilities. It is not possible to predict the outcome of any experiment that deals with a small number of particles or atoms: It is only possible to say that when there is a large number of them, they have certain kinds of average behavior.

A few examples should make this idea clearer. First we will look at the behavior of a macroscopic body that is not governed by quantum mechanics, a planet. The position of a planet can, in principle, be predicted with perfect accuracy. The only limiting factors are the size and speed of our computers and the accuracy of previous observations. Given the data and unlimited time for computation, there is nothing to stop us from calculating the position of Mars at 9:00 A.M. on April 14 one million years from now. In practice this would not be easy to do, but only because we do not know Mars's present position with perfect accuracy, and because we are not entirely sure about the size of the perturbations that would be induced by the other planets, such as Jupiter. If we cannot obtain an exact result, this is not due to any imperfections in the theory.

The decay of a radioactive atom, on the other hand, is an entirely different matter. There is no way that we can predict when any given atom will emit an alpha or a beta particle. It might happen five minutes from now, or only after the passage of a million years. All that quantum mechanics allows us to say is that if we have a large enough number of atoms, then approximately half of them will decay in a certain period of time.

This is the meaning of the commonly used term *half life*. Uranium 238, for example, decays to thorium 234 when it emits an alpha particle. The half life of the U-238 is 4.5 billion years, but this is only a statistical average. We simply cannot make any predictions about any individual atom. Events on this level are ruled by chance.

Numerous books have been written about the apparent failure of causality in quantum processes, and about the supposed philosophical implications of this fact. Numerous other books have been written to support the minority belief that it might be possible to interpret quantum mechanics in such a way that it is deterministic after all. The general feeling among scientists seems to be that the latter attempts have been unsatisfactory. Many of them think that it is better to give up the notion that subatomic behavior is deterministic than it is to complicate a very workable theory.

Everett's many-worlds theory is one way of putting determinism back into quantum mechanics. Everett has pointed out that it is not necessary to assume that the universe is governed by chance if we only allow ourselves to admit that there might be not just one universe, but an infinite number.

In Everett's theory, our universe splits off into innumerable replicas of itself at every moment. Each of these new universes splits again whenever a chance event takes place on the subatomic level. According to the theory, there exists an infinity of universes, each of which is destined to produce an infinite number of copies of itself.

To see how this happens, let us consider the uranium 238 atom. In the orthodox version of quantum mechanics, it may decay at any given moment in time. In Everett's theory, it decays at *every* moment of time, and each one of these events corresponds to a different universe. Meanwhile, similar bifurcations are taking place for every other uranium 238 atom in the universe, and every other object that is governed by quantum laws. Every electron, every proton, every neutrino causes the universe to split in two over and over again every time that it encounters a situation in which it has to make a "choice."

If Everett's theory is true, it is still impossible to say when an atom of uranium 238 is going to decay. The best we can do is to look at an atom that already has decayed and say that the decay took place at the only possible time that it could have *in our branch of the universe.*

Accepting the theory does not give us a very satisfactory kind of causality. Paradoxically, the many-worlds interpretation seems to simultaneously imply another kind of causality that seems *too* strong. For if we accept the theory then we must believe that quantum events that take place in galaxies billions of light years away cause us to split into innumerable replicas of ourselves at every moment. After all, whenever the universe splits, we must bifurcate too.

It would seem that such a fantastic idea would be easy to disprove. However, this is not the case; Everett's theory is mathematically identical to the standard version of quantum mechanics. This means that it is not possible to devise an experiment that could distinguish between the two. Furthermore, there seems to be no way to avoid the rather paradoxical conclusion that every experiment that confirms the standard quantum mechanics must be considered to be an experimental confirmation of Everett's theory also. Since the theories are mathematically identical, they make the same predictions.

Metaphysics is generally not considered to be part of science. The distinction between the two goes back at least as far as Aristotle (who seems to have coined the term; it means "after physics"). But as soon as we begin to consider something like the many-worlds theory, metaphysical questions begin to intrude. One cannot help wondering what the implications of such a theory would be for the doctrine of free will, or exactly what one is to make of the fact that it and standard quantum mechanics are identical. And of course the theory is based on a philosophical notion to begin with, namely that happenings on the subatomic level are no more governed by chance than are events in the world around us.

THE FATE OF THE UNIVERSE

Is there any connection between the many-universes theory of Everett and the infinite number of universes that some astronomers have postulated? There do seem to be some similarities. But there are many more differences. The splittings that Everett postulates do not seem to produce those universes that are lifeless because the expansion out of the big bang was too rapid, or because electromagnetism is related to some number other than $1/137$. Most of Everett's infinite universes would be just as full of life as this one is.

But it is interesting that both speculation about events that take place on the microscopic level and speculation about the universe as a whole should have led to many-universes theories. There may be more of a connection between microcosmos and macrocosmos than we believe.

If many different universes exist, it is not necessary to assume that they must be "parallel" in the sense that they exist simultaneously in time. It is not even certain that the word "simultaneously" has any meaning in this case. Each universe would have four dimensions: three of space and one of time. It could very well be that there was no connection between time in one universe and time in any of the others.

Or the universes could simply follow one another in time. If the universe really does oscillate, there could have been an infinite number of cycles in the past. Similarly, there may be an infinite number in the future. And if the universe "reprocesses" itself during each collapse the way Wheeler thinks it might, the majority of the infinite string of universes that are thereby produced might very well be lifeless. In the superspace theory, physical laws and constants change. As a result, there might be numerous universes like ours in which the fine structure constant is 1/137, and even more where it takes on a variety of different values.

If the universe oscillates an infinite number of times, every possible variation will eventually take place. This is true whether there is any reprocessing or not. In the latter case the range of possibilities is more limited. However, they seem to be varied enough to satisfy any but the most insatiable imagination. There would be universes in which the American Revolution never took place, and universes in which the electric light was invented by Socrates. There would be universes in which the reptiles, rather than the mammals, of our planet were the creatures that developed intelligence. And there would be universes in which the Milky Way galaxy did not exist at all. To be sure, there would be others which differed from ours only in insignificant ways. In some of them, the only difference might be that the Eiffel Tower was a few centimeters taller, or that you were given a different middle name, or that one of those uranium 238 atoms picked a different moment to decay.

Physicists generally like to believe that if two theories give an explanation of the same phenomena, then it is the simpler of the two theories that has the better chance of being true. I am not sure whether this is really a statement about the "simplicity" of nature, however. It may be nothing more than a reflection of the limitations of the human mind. Even supposedly "simple" theories such as general relativity (it is simple in the sense that it is based on very few assumptions) can produce equations so complex that no one can find a way to solve them. A really complicated theory could be more than humans could contend with.

Nevertheless, it would be interesting to apply the principle of simplicity to the many-universes concept. And it appears to me that the Peebles-Dicke hypothesis requires the smallest number of assumptions. Their reproducing-universe theory seems only to depend on the assumption that a black hole is a universe in miniature and that it can grow in mass every time it collapses and re-explodes into a new cycle.

Unfortunately, there appears to be no way to test this theory. In that respect it is in a class with Everett's many-worlds interpretation of quantum mechanics.

It is certainly possible to speculate about what this peculiar non-testability may imply, and I propose to do just that in the next chapter.

CHAPTER 9

Einstein's God,
Gödel's Proof,
St. Augustine's Curse

WHAT are we to make of all this speculation? Are there really other universes? If so, can they be reached through black hole gateways? Or are black holes, as Dicke and Peebles suggest, the seeds of new universes? Does the universe destroy itself over and over again in a catastrophic collapse, only to be reborn in a succession of big bangs? Is there anything in Wheeler's idea that the universe reprocesses itself in the big crunch so that the laws of nature are different every time there is a new big bang? Is this idea even consistent with Wheeler's other notion, that only universes that can give rise to life will come into existence?

At present, no observational evidence is capable of telling us whether any of these ideas are true or false. All are very difficult to test experimentally. It may not even be possible to determine whether the universe is open or closed. A half century of attempts to find out have established only that it is very close to the borderline.

The experiments that will be performed during the 1980s may settle the matter. If the neutrino turns out not to have mass, then we will be able to reasonably conclude that the universe is open. But suppose neutrinos turn out to have just enough mass—as they very well might—to provide the critical density. Then where will we be?

We may very well live in a closed, finite universe. But if we do, it is not possible to say what will happen after the big crunch. The idea that there will be a new big bang is one that scientists and lay people alike find very appealing. However, liking an idea is not sufficient to make it true. It is necessary to have evidence.

If I have belabored this point, it is a reaction to certain characteristics of the age in which we live. Today all sorts of things for which there is no convincing evidence are believed because people want to believe in them. I am thinking not only of such things as UFO's, the lost continent Atlantis, pyramid power, chariots of the gods, and the alleged discovery of Noah's Ark on Mt. Ararat, but also of some of the material that appears in supposedly scientific books.

Too often no attempt is made to distinguish between fact and speculation. At times facts are ignored if bringing them into the discussion would make an account of this theory or that one seem less interesting. For example, one book after another has made mention of the concept of black hole gateways without mentioning the fact that this subject lies on the fringes of scientific speculation. As a result, the public adds this concept to its store of "knowledge."

And yet astronomers are not even completely certain that black holes really exist. Black holes are a prediction of the theory of general relativity, and while the vast majority of astronomers feel that an excellent case has been made for the existence of a black hole in Cygnus X-1, the evidence that there really is a black hole there is of an indirect nature, and there are still a few doubters. If Cygnus X-1 turns out not to contain a black hole, then it will not be possible to say that there is any good evidence that black holes exist at all. If it is discovered that astronomers have interpreted the evidence incorrectly in this case, it will be hard for them to convince anyone that they have found a black hole somewhere else.

If one wants to know what the universe is like, it is necessary first to find out what might be true. Only after one has explored all the possibilities can one begin to make choices among them. Theoretical scientists invent imaginary universes because they want to know what our universe is like. Such an endeavor often contains elements of intellectual play.

To say this is not to discredit such activity. Even Newton recognized that theoretical work in physics often has a game-like character. "To myself I seem to have been only like a boy playing on the seashore," he said. But of course

Newton also knew how large the unknown was; he continued: "and diverting myself in now and then finding a smoother pebble or a prettier shell than ordinary, whilst the great ocean of truth lay all undiscovered before me."

If science is to explore the depths of this ocean of truth, then it is necessary to consider "crazy" ideas as well as "reasonable" ones. And if it encounters concepts that seem contrary to common sense, then perhaps it is common sense that will be discarded. Our ideas of what is and what is not reasonable are derived from our experiences with ordinary objects in the everyday world. There is no reason to assume that our very limited common-sense ideas are necessarily applicable to a universe that is 18 billion years old, and which very well might stretch out to infinity.

But one is not free to imagine anything that one likes; there are some very real constraints. A theory cannot disregard observed facts. For example, one cannot build a cosmological theory on the assumption that the universe is 5 percent helium; observations indicate that the true figure is 25 percent. Similarly, one cannot ignore the existence of the cosmic microwave background, which is very obviously there.

One other important requirement must be met before a theory can be taken seriously: It must not seem too contrived. Scientists distrust explanations that are too complicated. After all, it is possible to "prove" almost anything by piling one ad hoc assumption on top of another. If such a thing were allowed, then even flat-earth theories would have to be taken seriously. All the flat-earth advocate would have to do would be to add another assumption every time an objection was made. If one is allowed to make enough assumptions, then anything can be explained—or explained away.

The rule that the simplest theory is always the best is, admittedly, a philosophical prejudice. But if one did not make certain philosophical assumptions (e.g., that the laws of nature are the same everywhere), it would not be possible to do physics at all. There is no way that anyone can prove that the universe does not resemble a Rube Goldberg invention. However, the assumption of simplicity has worked so well since the dawn of modern science that it is not likely to be abandoned now.

If a theory meets the above requirements—and certain other obvious ones, such as being possible to understand, and that it contain no mathematical mistakes—then the next step is to examine the predictions that it makes and compare these predictions to observations. The better the predictions are confirmed, the more established the theory will be. As we saw in Chapter 1, the reason why general relativity is considered to be so firmly established is that it predicts a number of different effects—such as the bending of light and of radio waves—which have been tested in experiments of very high accuracy. Similarly, the reason why physicists have such faith in quantum mechanics is the fact that it, too, has been subjected to experimental test so many times.

Unfortunately, it is not so easy to test some of the cosmological theories that have been described in this book. As I noted in the previous chapter, there do not seem to be any experiments that could be performed to determine whether the many-worlds interpretation of quantum mechanics is correct. It is this, more than its apparent "craziness," that causes scientists to view it with a certain amount of skepticism.

The Dicke-Peebles hypothesis that black holes might be the seeds of new universes appears to be just as difficult to confirm or disprove. We can neither see into black holes nor travel to other universes. As a result, it is impossible to say just how the theory could be tested.

Even the oscillating universe theory, which was originally proposed more than fifty years ago, does not seem to be subject to experimental test. After all, we cannot look into the future to see what will happen after the universe collapses. And when we look very far into the past, all we see is the microwave radiation that is a remnant of the big bang.

Of course it is possible that we will find the answers at some indefinite time in the future. During the twentieth century, science has made discoveries at such a rate that our conception of the universe has been transformed many times over. If there are questions that can never be answered, one hesitates to make any statements about what they are. It is well to remember that in the nineteenth century, French mathematician and philosopher Auguste Comte gave the chemical composition of stars as an example of unattainable knowledge. Today the makeup of stars can be studied very easily. One need only analyze the spectrum of the light a star emits.

Since the times of Kepler and of Newton, scientists have tended to believe that most questions would eventually be answered. And of course the successes achieved by modern science have done nothing to disparage this belief.

Einstein expressed this idea in an especially poetic way: "Raffiniert ist der Herrgott, aber böshaft ist er nicht." God is subtle, but not malicious.

As we know, Einstein was not right about everything. In fact, he made quite a number of mistakes during his illustrious scientific career. Therefore it is not unreasonable to ask whether we really will solve all of the problems that Einstein's God has posed. Are the laws of nature fundamentally knowable? Or is science, in the field of cosmology at least, beginning to encounter barriers beyond which human understanding cannot pass? Must it be possible to answer any question that we are capable of asking? Or could it be that there are limits to scientific knowledge?

It might be possible to gain some insight into the nature of the problem by turning away from cosmology for a moment, and looking at the current situation in mathematics.

Mathematics has its unanswered questions too. But in this case, matters are not complicated by the need to amass observational data. When there exists a mathematical question that has not been answered, the failure to find a solution cannot be attributed to the fact that the right experiments have not yet been performed; when mathematical problems persist, this is caused by the fact that the best minds have not been able to solve them.

There are numerous unsolved problems in mathematics. One of the best known concerns a conjecture that was made by the eighteenth-century German amateur mathematician Christian Goldbach. Not surprisingly, it is known as Goldbach's conjecture.

Goldbach happened to notice that every even number is apparently the sum of two primes.* For example:

$$4 = 3 + 1$$
$$6 = 5 + 1$$
$$28 = 23 + 5$$
$$100 = 47 + 53$$

Goldbach could discover no exception to this rule. Every even number that he could think of was expressible as the sum of two primes in at least one way (and of course some of them could be so expressed in more than one way; for instance, 28 is also 11 + 17, etc). One would expect that such a simple and apparently true idea should be easy to prove as a mathematical theorem. But Goldbach could not find a way to do it.

Goldbach wrote a letter describing the problem to Swiss mathematician Leonhard Euler. At the time, Euler was one of the most prominent mathematicians in Europe; today he is regarded as one of the greatest mathematicians of all time.

But Euler could not prove the truth of the conjecture either. Neither was he able to disprove it (to do this, he would only have had to discover a counterexample; i.e., a single case in which the conjecture was not true). Goldbach's conjecture seemed to be true, but Euler could think of no way that this could be demonstrated.

And this is where things stand today. Modern mathematicians are no nearer a proof than Goldbach and Euler were. The best that they have been able to do is to show

* A *prime* is a number that is divisible only by itself and by 1. For example, 1, 2, 3, 5, 7, and 11 are primes. But 4 (which is equal to 2 × 2) and 6 (3 × 2) are not.

that any number—odd or even—can be expressed as the sum of no more than 300,000 primes. A proof of the simple observation that every even number can be expressed as two primes continues to elude them.

It would certainly be possible to program a computer to test each even number, one by one, to see whether or not the conjecture held. But this would not constitute a mathematical proof. A theorem is something which demonstrates that a statement is true in every case. A computer could show that it was true only in every case that had so far been considered. Even if it tested every number up to 1,000,000,000,000,000,000, it would not have shown that the conjecture was valid for 1,000,000,000,000,000,002.

Most mathematicians believe that Goldbach's conjecture is almost certainly true. But they do not waste much time trying to prove it. Many of them suspect, in fact, that it is a statement that is impossible to prove.

Since 1931 it has been known that mathematical systems can contain *undecidable statements*, propositions that are true but that cannot be proved. It is thought that Goldbach's conjecture may be an example of such a statement.

The history of mathematics stretches thousands of years into the past. During most of this time, it has been thought of as the one subject in which things could be known with absolute certainty. Either a proposition was true or it was false. It was one or the other; there was no middle ground. Furthermore, once a theorem was proved, it was proved forever.

There have always been unsolved problems. Until fairly recently mathematicians were confident that these would eventually be disposed of. Indeed, this is exactly what happened in many cases.

And then in 1931 a young mathematician at the University of Vienna named Kurt Gödel shattered that complacency by publishing a paper which challenged the very foundations of mathematics. Today, the contents of that paper are commonly referred to as *Gödel's proof.*

Perhaps "Gödel's proofs" would be a more accurate term, since Gödel proved not one statement, but two. He showed, first, that no mathematical system that was at least as complex as ordinary arithmetic could ever be proved to be consistent. That is, one could not be certain that the system would not eventually produce theorems that would contradict one another. And of course if something like that did happen, the whole mathematical edifice would fall to the ground. One simply cannot place any trust in a mathematics that contains inconsistencies; it is one of the rules of logic that two statements that contradict one another can be used to prove anything at all, if one assumes that both are true.

It was the next part of Gödel's proof that introduced the concept of undecidable statements. In the second part of his paper, Gödel showed that there had to exist propositions that were true but could not be shown to be true. Gödel did not say what they were. It so happens that demonstrating that any given statement is undecidable is just as impossible as proving it. So even though mathematicians think that Goldbach's conjecture is a proposition of this type, they will never be able to be certain.

If there is anything that is characteristic of us humans, it is our longing for certainty. Are we to be denied it even in mathematics? Could it turn out that 2 plus 2 does not necessarily equal 4? Will we be forced to give up the multiplication tables that we labored to memorize when we were in grade school?

As a matter of fact, consistency can be proven for systems that are simple enough (the one consisting of all the whole numbers from one to a billion would be one example). It is only when the number of elements in a system becomes infinite that problems begin to arise.

Furthermore, contradictions have not been discovered in the types of mathematics used by physics, such as calculus and the theory of differential equations. At this point, the physical sciences seem not to have been much affected by the discovery that Gödel made.

However, the very fact that there are limits on what can be known in mathematics leads one to wonder whether there might not also be limits in physics. If there are, it seems reasonable that one would be most likely to encounter them in fields that, like cosmology, lie at the boundaries of human speculation. After all, mathematics is the language of physics. If there are uncertainties in the former field, there should be at least as many in the latter.

One is tempted to ask whether certain hypotheses concerning the things that take place inside black holes, or the events that happen after the big crunch, might not be analogous to undecidable statements in mathematics. But of course this is something about which we cannot be sure. Even in mathematics one does not know what the undecidable statements are.

There may or may not be questions about the universe that cannot be answered. We only know that there are some that have not been answered yet, although they have been asked in one age after another. These questions, which are frequently the most baffling, seem also to be the most intriguing and the most fundamental. Even though some of them have puzzled us for centuries, we persist in looking for answers.

In the fourth century A.D., St. Augustine was asked, "What was God doing before He made the heaven and the earth?".

Now Augustine did not, as some authors state, reply that God was preparing hell for those who pried into such matters. Augustine does mention this answer in his *Confessions*, but he goes on to say that he would not give so frivolous an answer.* However, I don't think that I can be blamed if I refer to the joking reply as "St. Augustine's curse." Augustine says that he did not give this answer; nevertheless he did record it.

What did happen before the Creation? Sixteen hundred years have passed since Augustine was asked that, and we still do not know. It may be that nothing did. It may be that, as the ancient Stoics and those astronomers who favor the oscillating universe theory believe, there were many universes which preceded ours.

During the closing years of the nineteenth century, physicists thought that physical science was practically complete, and that all that remained to be done was to fill out the corners of scientific knowledge. Slight modifications would be made in existing theories, and experiments would be conducted to measure various quantities to the next decimal place, but they did not believe that any major

* The answer that he does give is strikingly similar to that suggested by some contemporary physicists. He says that time did not exist before the Creation.

changes in their conception of the universe were imminent.

We all know what happened next. Within a few years, physics was in upheaval. One scientific revolution followed another. Scientists found that they could hardly absorb all of the implications of one discovery before a half dozen others were made. Radioactivity was discovered in 1895, quantum theory (the forerunner of quantum mechanics) in 1901, special relativity in 1905, the nuclear atom in 1911, and general relativity in 1915. And that was only the beginning.

Today we realize that there will always be questions. Contrary to what the late-nineteenth-century physicists thought, nature is not something that can be completely understood with the help of a handful of theories. On the contrary, nature is infinite in its variety. The more scientific knowledge increases, the more we realize there is much we do not know.

There may even be some things that we cannot know. But if there are, it is not certain that we can tell what they are—at least not until some mathematician of the future tells us how to recognize undecidable statements.

So even if it does turn out that there are things that cannot be known, we are not likely to stop asking questions. Not even St. Augustine was willing to threaten us with hell for doing that.

BIBLIOGRAPHY

Aaronson, Marc; Huchra, John; and Mould, Jeremy. "The Infrared Luminos-
   ity/H_1 Velocity–Width Relation and Its Application to the Distance Scale."
   Astrophysical Journal, Vol. 229 (1979), pp. 1–13.
Aaronson, Marc; Mould, Jeremy; and Huchra, John. "A Distance Scale from
   the Infrared Magnitude /H_1 Velocity–Width Relation. I. The Calibration."
   Astrophysical Journal, Vol. 237 (1980), pp. 655–665.
Asimov, Isaac. *The Neutrino*. Garden City, New York: Doubleday, 1966.
Augustine, Saint. *Confessions of St. Augustine*. Garden City, New York:
   Doubleday Image, 1960.
Bahcall, John N., and Davis, Raymond, Jr. "Solar Neutrinos: A Scientific Puz-
   zle." *Science*, Vol. 191 (1976), pp. 264–267.
Barrow, John D., and Tipler, Frank J. "Eternity Is Unstable." *Nature*, Vol. 276
   (1978), pp. 453–459.
Bartusiak, Marcia F. "Experimental Relativity: Its Day in the Sun." *Science
   News*, Vol. 116 (1979), pp. 140–142.
Bernstein, Jeremy. *Einstein*. New York: Viking, 1973.
Berry, Michael. *Principles of Cosmology and Gravitation*. Cambridge: Cam-
   bridge University Press, 1976.
Bok, Bart J. "The Milky Way Galaxy." *Scientific American*, Vol. 244, No. 3
   (March, 1981), pp. 92–120.
Calder, Nigel. *Einstein's Universe*. New York: Viking, 1979.

Bibliography

——. *The Key to the Universe.* New York: Viking, 1977.

Canuto, V., and Hsieh, S. H. "Case for an Open Universe." *Physical Review Letters,* Vol. 44 (1980), pp. 695–698.

Chaffee, Frederic H., Jr. "The Discovery of a Gravitational Lens." *Scientific American,* Vol. 243, No. 5 (November, 1980), pp. 70–78.

Close, Frank E. "Weighing Neutrinos: A Small Wait for a Small Weight?" *Nature,* Vol. 289 (1981), pp. 747–748.

Cocke, W. J. "Statistical Time Symmetry and Two-Time Boundary Conditions in Physics and Cosmology." *Physical Review,* Vol. 160 (1967), pp. 1165–1170.

Collins, C. B., and Hawking, S. W. "Why Is the Universe Isotropic?" *Astrophysical Journal,* Vol. 180 (1973), pp. 317–334.

Davies, P. C. W. "Closed Time as an Explanation of the Black Body Background Radiation." *Nature Physical Science,* Vol. 240 (1972), pp. 3–5.

——. *The Forces of Nature.* Cambridge: Cambridge University Press, 1979.

——. "How Special Is the Universe?" *Nature,* Vol. 249 (1974), pp. 208–209.

——. *The Physics of Time Asymmetry.* Berkeley: University of California Press, 1977.

——. *The Runaway Universe.* New York: Harper & Row, 1978.

Davis, Marc, et al. "On the Virgo Supercluster and the Mean Mass Density of the Universe." *Astrophysical Journal Letters,* Vol. 238 (1980), pp. L113–116.

De Rújula, A., and Glashow, S. L. "Galactic Neutrinos and uv Astronomy." *Physical Review Letters,* Vol. 45 (1980), pp. 942–944.

——. "Neutrino Weight Watching." *Nature,* Vol. 286 (1980), pp. 755–756.

"Do Neutrinos Oscillate from One Variety to Another?" *Physics Today,* Vol. 33 (July, 1980), pp. 17–19.

Dyson, Freeman. *Disturbing the Universe.* New York: Harper & Row, 1979.

Edmunds, M. G. "Open Debate." *Nature,* Vol. 288 (1980), pp. 431–432.

Ellis, G. F. R. "Limits to Verification in Cosmology." *Annals of the New York Academy of Sciences,* Vol. 336 (1980), pp. 130–160.

Faber, S. M., and Gallagher, J. S. "Masses and Mass-to-Light Ratios of Galaxies." *Annual Review of Astronomy and Astrophysics,* Vol. 17 (1979), pp. 135–187.

Ferris, Timothy. *The Red Limit.* New York: Morrow, 1977.

Feynman, Richard. *The Character of Physical Law.* Cambridge, Massachusetts: M.I.T. Press, 1967.

"From Russia with Mass: Neutrinos." *Science News,* Vol. 118 (1980), pp. 228–229.

Fuller, Robert W., and Wheeler, John A. "Causality and Multiply Connected Space-Time." *Physical Reivew,* Vol. 128 (1962), pp. 919–929.

Gamow, George. *My World Line.* New York: Viking, 1970.

Gardner, Martin. *The Ambidextrous Universe.* 2nd Ed. New York: Scribner's, 1979.

Bibliography

Giacconi, Ricardo. "The Einstein X-Ray Observatory." *Scientific American,* Vol. 242, No. 2 (February, 1980), pp. 80–102.

Giacconi, Ricardo, and Tananbaum, Harvey. "The Einstein Observatory: New Perspectives in Astronomy." *Science,* Vol. 209 (1980), pp. 865–876.

Gold, Thomas. "Relativity and Time." In *The Encyclopedia of Ignorance,* edited by Ronald Duncan and Miranda Weston-Smith. Oxford: Permagon Press, 1977.

Goldhaber, M.; Langacker, P.; and Slansky, R. "Is the Proton Stable?" *Science,* Vol. 210 (1980), pp. 851–860.

Gorenstein, Paul, and Tucker, Wallace. "Rich Clusters of Galaxies." *Scientific American,* Vol. 239, No. 5 (November, 1978), pp. 110–128.

Gott, J. Richard, III, and Turner, Edwin L. "The Mean Luminosity and Mass Densities in the Universe." *Astrophysical Journal,* Vol. 209 (1976), pp. 1–5.

Gott, J. Richard, III, et al. "An Unbound Universe?" *Astrophysical Journal,* Vol. 194 (1974), pp. 543–553.

"Gravity's Lens: Squinting at a Galaxy." *Science News,* Vol. 117 (1980), pp. 36–37.

Gribbin, John. *Galaxy Formation.* New York: Wiley, 1976.

———. *White Holes.* New York: Delacorte, 1977.

Hartline, Beverly Karplus. "Double Hubble, Age in Trouble." *Science,* Vol. 207 (1980), pp. 167–169.

Hartman, William K. *Astronomy: The Cosmic Journey.* Belmont, California: Wadsworth, 1978.

Hawking, S. W., and Ellis, G. F. R. *The Large Scale Structure of Space-Time.* Cambridge: Cambridge University Press, 1973.

Hawking, S. W., and Israel, W., eds. *General Relativity.* Cambridge: Cambridge University Press, 1979.

"Holes in History." *Scientific American,* Vol. 241, No. 1 (July, 1979), p. 80.

Hoyle, Fred. *Astronomy and Cosmology.* San Francisco: Freeman 1975.

———. *Ten Faces of the Universe.* San Francisco: Freeman, 1977.

Hoyle, Fred, and Narlikar, Jayant. *The Physics-Astronomy Frontier.* San Francisco: Freeman, 1980.

Hubble, Edwin. *The Realm of the Nebulae.* New Haven: Yale University Press, 1936.

———. "A Relation Between Distance and Radial Velocity among Extra-Galactic Nebulae." *Proceedings of the National Academy of Science,* Vol. 15 (1929), pp. 168–173.

Huchra, John. "The Cosmic Calendar." Preprint, Harvard-Smithsonian Center for Astrophysics.

———. "The Extragalactic Distance Scale and the Age of the Universe." Preprint, Harvard-Smithsonian Center for Astrophysics.

"Identity Crisis." *Scientific American,* Vol. 243, No. 1 (July, 1980), pp. 72–75.

John, Laurie, ed. *Cosmology Now.* New York: Taplinger, 1976.

Kaufmann, William J., III. *The Cosmic Frontiers of General Relativity.* Boston: Little, Brown, 1977.

Bibliography

Kazanas, Demosthenes; Schramm, David N.; and Hainebach, Kem. "A Consistent Age for the Universe." *Nature*, Vol. 274 (1978), pp. 672–673.

Kline, Morris. *Mathematics: The Loss of Certainty.* New York: Oxford University Press, 1980.

Kristian, Jerome; Sandage, Allan; and Westphal, James A. "The Extension of the Hubble Diagram. II. New Redshifts and Photometry of Very Distant Galaxy Clusters: First Indication of a Deviation of the Hubble Diagram from a Straight Line." *Astrophysical Journal*, Vol. 221 (1978), pp. 383–394.

Landsberg, Peter T., and Evans, David A. *Mathematical Cosmology.* Oxford: Clarendon Press, 1977.

Landsberg, Peter T., and Park, D. "Entropy in an Oscillating Universe." *Proceedings of the Royal Society of London*, Series A, Vol. 346 (1975), pp. 485–495.

Lebovitz, Norman R.; Reid, William H.; and Vandervoort, Peter O. *Theoretical Perspectives in Astrophysics and Relativity.* Chicago: University of Chicago Press, 1978.

Maffei, Paolo. *Monsters in the Sky.* Cambridge, Massachusetts: M.I.T. Press, 1980.

"Massive Stellar Mass." *Science News*, Vol. 119 (1981), p. 36.

Meier, David L., and Sunyaev, Rashid L. "Primeval Galaxies." *Scientific American*, Vol. 241, No. 5 (November, 1979), pp. 130–144.

Misner, Charles W.; Thorne, Kip S.; and Wheeler, John Archibald. *Gravitation.* San Francisco: Freeman, 1973.

Mitton, Simon. *Exploring the Galaxies.* New York: Scribner's, 1976.

Morris, Richard. *The End of the World.* Garden City, New York: Doubleday Anchor, 1980.

———. *Light.* Indianapolis: Bobbs-Merrill, 1979.

Motz, Lloyd. *The Universe.* New York: Scribner's, 1975.

Noble, R. G., and Walsh, D. "MTRLI Observations of the Double QSO at 408 MHz." *Nature*, Vol. 288 (1980), pp. 69–70.

North, J. D. *The Measure of the Universe.* Oxford: Clarendon Press, 1965.

"An Optical Three-Way Split." *Science News*, Vol. 118 (1980), p. 4.

Pathria, R. K. "The Universe as a Black Hole." *Nature*, Vol. 240 (1972), pp. 298–299.

Peebles, P. J. E. *The Large-Scale Structure of the Universe.* Princeton: Princeton University Press, 1980.

———. *Physical Cosmology.* Princeton: Princeton University Press, 1971.

Perl, Martin L. "The Tau Lepton." *Annual Review of Nuclear Particle Science*, Vol. 30 (1980), pp. 299–335.

Pines, David. "Accreting Neutron Stars, Black Holes and Degenerate Dwarf Stars." *Science*, Vol. 207 (1980), pp. 597–606.

Plato. *The Collected Dialogues.* Princeton: Princeton University Press, 1961.

"The Question of a Quintuple Quasar." *Science News*, Vol. 118 (1980), p. 36.

Raychaudhuri, A. K. *Theoretical Cosmology.* Oxford: Clarendon Press, 1979.

Reines, Frederick. "The Early Days of Experimental Neutrino Physics." *Science*, Vol. 203 (1979), pp. 11–16.

Bibliography

Rowan-Robinson, Michael. *Cosmology*. Oxford: Clarendon Press, 1977.

Rubin, Vera C.; Ford, W. Kent; and Thonnard, Norbert. "Rotational Properties of 21 Sc Galaxies with a Large Range of Luminosities and Radii, from NGC 4605 (R = 4 kpc) to UGC 2885 (R = 122 kpc)." *Astrophysical Journal*, Vol. 238 (1980), pp. 471–487.

Sandage, Allan. "The Ability of the 200-inch Telescope to Discriminate Between Selected World Models." *Astrophysical Journal*, Vol. 133 (1961), pp. 355–392.

———. "Cosmology: A Search for Two Numbers." *Physics Today*, Vol. 23, No. 2 (1970), pp. 34–41.

———. "Distances to Galaxies: The Hubble Constant, the Friedmann Time, and the Edge of the World." *Quarterly Journal of the Royal Astronomical Society*, Vol. 13 (1972), pp. 282–296.

Sandage, Allan, and Tammann, G. A. "Steps Toward the Hubble Constant. VII. Distances to NGC 2403, M 101 and the Virgo Cluster Using 21 Centimeter Line Widths Compared with Optical Methods: The Global Value of H_0." *Astrophysical Journal*, Vol. 210 (1976), pp. 7–24.

Schramm, David N. "Neutrino Astronomy." *Annals of the New York Academy of Sciences*, Vol. 336 (1980), pp. 380–388.

Schramm, David N., and Wagoner, Robert V. "Element Production in the Early Universe." *Annual Review of Nuclear Science*, Vol. 27 (1977), pp. 37–74.

Scientific American editors. *Cosmology + 1*. San Francisco: Freeman, 1977.

———. *New Frontiers in Astronomy*. San Francisco: Freeman, 1975.

Segrè, Emilio. *From X-Rays to Quarks*. San Francisco: Freeman, 1980.

Shipman, Harry L. *Black Holes, Quasars and the Universe*. 2nd Ed. Boston: Houghton Mifflin, 1980.

———. *The Restless Universe*. Boston: Houghton Mifflin, 1978.

Silk, Joseph. *The Big Bang*. San Francisco: Freeman, 1980.

———. "The Intergalactic Medium." *Nature*, Vol. 292 (1981), p. 83.

Singh, Jagjit. *Great Ideas and Theories of Modern Cosmology*. 2nd Ed. New York: Dover, 1970.

Stecker, F. W. "Have Massive Cosmological Neutrinos Already Been Detected?" *Physical Review Letters*, Vol. 45 (1980), pp. 1460–1462.

Sullivan, Walter. *Black Holes*. Garden City, New York: Doubleday Anchor, 1979.

Sunyaev, R. A., and Zel'dovich, Ya. B. "Microwave Background Radiation as a Probe of the Contemporary Structure and History of the Universe." *Annual Review of Astronomy and Astrophysics*, Vol. 18 (1980), pp. 537–560.

Symbalisty, E. M. D.; Yang, J.; and Schramm, D. N. "Neutrinos and the Age of the Universe." *Nature*, Vol. 288 (1980), pp. 143–145.

Tayler, R. J. "Neutrino Stability and Cosmological Helium Production." *Nature*, Vol. 274 (1978), pp. 232–234.

Thomsen, Dietrick. E. "Shaking Neutrino Mass." *Science News*, Vol. 118 (1980), pp. 298–300.

Bibliography

————. "Ups and Downs of Neutrino Oscillation." *Science News,* Vol. 117 (1980), pp. 377–383.

————. "X-ray Astronomy Comes of Age." *Science News,* Vol. 115 (1979), pp. 234–235.

Tinsley, Beatrice M. "Accelerating Universe Revisited." *Nature,* Vol. 273 (1978), pp. 208–11.

————. "The Cosmological Constant and Cosmological Change." *Physics Today,* Vol. 30 (June, 1977), pp. 32–38.

Tremaine, Scott, and Gunn, James E. "Dynamical Role of Light Neutral Leptons in Cosmology." *Physical Review Letters,* Vol. 42 (1979), pp. 407–409.

Turner, Michael S., and Schramm, David N. "Cosmology and Elementary-Particle Physics." *Physics Today,* Vol. 32, No. 9 (September, 1979), pp. 42–48.

Wald, Robert M. *Space, Time, and Gravity.* Chicago: University of Chicago Press, 1977.

Waldrop, M. Mitchell. "Massive Neutrinos: Masters of the Universe?" *Science,* Vol. 211 (1980), pp. 470–472.

"Weighed in the Balance and Found: Neutrino." *Science News,* Vol. 117 (1980), pp. 292–293.

Weinberg, Steven. *The First Three Minutes.* New York: Basic Books, 1977.

"Whatever Happened to Neutrino Mass?" *Nature,* Vol. 287 (1980), p. 481.

Wheeler, John Archibald. "Superspace." In *Analytical Methods in Mathematical Physics,* edited by Robert P. Gilbert and Roger G. Newton. New York: Gordon and Breach, 1970.

Wheeler, John Archibald, and Patton, C. M. "Is Physics Legislated by Cosmology?" In *The Encyclopedia of Ignorance,* edited by Ronald Duncan and Miranda Weston-Smith. Oxford: Permagon Press, 1977.

Wilczek, Frank. "The Cosmic Asymmetry between Matter and Antimatter." *Scientific American,* Vol. 243, No. 6 (December, 1980), pp. 82–90.

Yang, Jongmann, et al. "Constraints on Cosmology and Neutrino Physics from Big Bang Nucleosynthesis." *Astrophysical Journal,* Vol. 227 (1979), pp. 697–704.